Advances in Intelligent Systems and Computing

Volume 1276

The series "Advances in Intelligent Systems and Computing" contains publications on theory, applications, and design methods of Intelligent Systems and Intelligent Computing. Virtually all disciplines such as engineering, natural sciences, computer and information science, ICT, economics, business, e-commerce, environment, healthcare, life science are covered. The list of topics spans all the areas of modern intelligent systems and computing such as: computational intelligence, soft computing including neural networks, fuzzy systems, evolutionary computing and the fusion of these paradigms, social intelligence, ambient intelligence, computational neuroscience, artificial life, virtual worlds and society, cognitive science and systems, Perception and Vision, DNA and immune based systems, self-organizing and adaptive systems, e-Learning and teaching, human-centered and human-centric computing, recommender systems, intelligent control, robotics and mechatronics including human-machine teaming, knowledge-based paradigms, learning paradigms, machine ethics, intelligent data analysis, knowledge management, intelligent agents, intelligent decision making and support, intelligent network security, trust management, interactive entertainment, Web intelligence and multimedia.

The publications within "Advances in Intelligent Systems and Computing" are primarily proceedings of important conferences, symposia and congresses. They cover significant recent developments in the field, both of a foundational and applicable character. An important characteristic feature of the series is the short publication time and world-wide distribution. This permits a rapid and broad dissemination of research results.

Indexed by SCOPUS, DBLP, EI Compendex, INSPEC, WTI Frankfurt eG, zbMATH, Japanese Science and Technology Agency (JST), SCImago.

More information about this series at http://www.springer.com/series/11156

Jyotsna Kumar Mandal · Imon Mukherjee ·
Sambit Bakshi · Sanjay Chatterji · Pankaj K. Sa
Editors

Computational Intelligence and Machine Learning

Proceedings of the 7th International Conference on Advanced Computing, Networking, and Informatics (ICACNI 2019)

Springer

Editors
Jyotsna Kumar Mandal
Department of Computer Science and Engineering
University of Kalyani
Kalyani, West Bengal, India

Sambit Bakshi
Department of Computer Science and Engineering
National Institute of Technology Rourkel
Rourkela, Odisha, India

Pankaj K. Sa
Department of Computer Science and Engineering
National Institute of Technology Rourkel
Rourkela, Odisha, India

Imon Mukherjee
Department of Computer Science and Engineering
Indian Institute of Information Technology Kalyani
Kalyani, West Bengal, India

Sanjay Chatterji
Department of Computer Science and Engineering
Indian Institute of Information Technology Kalyani
Kalyani, West Bengal, India

ISSN 2194-5357 ISSN 2194-5365 (electronic)
Advances in Intelligent Systems and Computing
ISBN 978-981-15-8609-5 ISBN 978-981-15-8610-1 (eBook)
https://doi.org/10.1007/978-981-15-8610-1

This Springer imprint is published by the registered company Springer Nature Singapore Pte Ltd.
The registered company address is: 152 Beach Road, #21-01/04 Gateway East, Singapore 189721, Singapore

Preface

The 7th International Conference on Advanced Computing, Networking, and Informatics (ICACNI, 2019) was held during 20–21 December 2019, at the Indian Institute of Information Technology Kalyani, India. There were theoretical research works from Machine Learning and other algorithm domains and their applications.

This book focuses on both theory and applications in the broad areas of computational intelligence and machine learning. ICACNI 2019 presents research papers in the areas of advanced computing, networking, and informatics. It brings together contributions from scientists, professors, scholars, and students and presents essential information on the topic. It also discusses the practical challenges encountered and the solutions used to overcome them, the goal being to promote the "translation" of basic research into applied research and of applied research into practice. The works presented here also demonstrate the importance of basic scientific research in a range of fields.

Some of the works are on Natural Language Processing domain like a sarcasm detection tool using Deep Neural Network. An article is on sentiment analysis using VADER for Twitter data. The book consists of an article on Latent Dirichlet Allocation (LDA)-based domain-specific sentiment analysis applicable for the Twitter dataset. Type classification problem is addressed in another article. A model to retrieve news articles related to the news currently being read based on their context and background information is also presented. The user-defined parameters for clustering algorithm are done in another article by harnessing the capabilities of Genetic Algorithm (GA).

There are a few articles based on image processing tasks also. Some other authors devised few techniques based on some circuit parameters. A numerical model of the mammalian ventricular cell to exhibit the effects of pacing and stimulus variations in the Luo–Rudy Phase I model is also presented in this book. Another work is on distributed algorithm for computational mobile entities, popularly known as swarm robots that operate and move in continuous space to arrange themselves in a straight line. A Deep Learning Model (DLSTM) is also presented to forecast the air pollution. A Canadian Middle Atmosphere Model (CMAM30) data is analyzed for the temperature variation from 1000 to 0.001 hPa from 2009 to 2010 to study the effect of window size on two-dimensional Fourier spectra.

An efficient adaptive Weighted Hybrid Queue Scheduling Scheme (WHQSS) is also contributed by the authors. An integrated improved hybrid cuckoo-GKO-based low power consumption routing protocol is suggested by another author. Convolutional Neural Network (CNN) model-based steganalysis is suggested by the authors. A novel watermarking-based approach is proposed by some authors to enable digital media assets to be maintained with their metadata persistently and robustly.

The editors would like to thank the authority of IIIT Kalyani, India, for providing infrastructural support to host the conference at IIIT Kalyani. Our sincere gratitude to the authorities of Springer Nature for publishing the presented papers of this conference as a book series.

This volume will be useful to the researchers, budding engineers, and undergraduate and postgraduate students of related disciplines.

Kalyani, India Jyotsna Kumar Mandal
Kalyani, India Imon Mukherjee
Rourkela, India Sambit Bakshi
Kalyani, India Sanjay Chatterji
Rourkela, India Pankaj K. Sa
Editors

Contents

About the Editors

Jyotsna Kumar Mandal, M.Tech. in Computer Science from the University of Calcutta in 1987, awarded Ph.D. (Engineering) in Computer Science and Engineering by Jadavpur University in 2000. Presently, he is working as Professor of Computer Science & Engineering, and former Dean, Faculty of Engineering, Technology & Management, KU, for two consecutive terms during 2008–2012. He is Director, IQAC, Kalyani University, and Chairman, CIRM, and Placement Cell. He served as Professor of Computer Applications, Kalyani Government Engineering College, for two years, as Associate Professor of Computer Science for eight years at North Bengal University, as Assistant Professor of Computer Science North Bengal University for seven years, and as Lecturer at NERIST, Itanagar, for one year. He has 32 years of teaching and research experience in coding theory, data and network security and authentication; remote sensing & GIS-based applications, data compression, error correction, visual cryptography, and steganography. He was awarded 23 Ph.D. degrees, one submitted and 8 are pursuing. He has supervised 03 M.Phil., more than 70 M.Tech., and more than 125 M.C.A dissertations. He is Guest Editor of MST Journal (SCI indexed) of Springer. He published more than 400 research articles out of which 170 articles in international journals. He published 7 books from LAP Germany, IGI Global, etc. He has organized 31 international conferences and Corresponding Editors of edited volumes and conference publications of Springer, IEEE, Elsevier, etc., and edited 32 volumes as Volume Editor.

Imon Mukherjee received his Ph.D. from Jadavpur University, Kolkata, India, in 2015, respectively. Currently, he is working as Faculty in-Charge, Academics, and an Assistant Professor (Grade-I) in Computer Science & Engineering, Indian Institute of Information Technology Kalyani, India (an Institute of National Importance under MHRD, Government of India). Earlier, he was an Assistant Professor in the Department of Computer Science and Engineering of the Institute of Technology & Marine Engineering, now The Neotia University, West Bengal, India, from August 2006 to December 2011. Then from January 2012, he was associated with the Department of Computer Science and Engineering, St. Thomas' College of Engineering & Technology, Kolkata, India, with the same designation. His research focuses on information security and data analytics. He is associated as Reviewer, PC member, and Track Session Chair of many reputed conferences and journals like IEEE Transactions on Circuits and Systems for Video Technology. He is currently supervising 5 Ph.D. students at IIIT Kalyani. He has acted as PI and Co-PI of many projects funded by DST, Government of West Bengal, etc.

Sambit Bakshi is currently with Centre for Computer Vision and Pattern Recognition of National Institute of Technology Rourkela, India. He also serves as an Assistant Professor in the Department of Computer Science & Engineering of the institute. He earned his Ph.D. degree in Computer Science & Engineering. His area of interest includes surveillance and biometric authentication. He is a senior member of IEEE since 2019. He currently serves as Associate Editor of International Journal of Biometrics (2013–), IEEE Access (2016–), Innovations in Systems and Software Engineering: A NASA Journal (2016–), Expert Systems (2017–), and Plos One (2017–). He has served/is serving as Guest Editor for reputed journals like Multimedia Tools and Applications, IEEE Access, Innovations in Systems and Software Engineering: A NASA Journal, Computers and Electrical Engineering, IET Biometrics, and ACM/Springer MONET. He is serving as the Vice-Chair for IEEE Computational Intelligence Society Technical Committee for Intelligent Systems Applications for the year 2019. He received the prestigious Innovative Student Projects Award 2011 from Indian National Academy of Engineering (INAE) for his master's thesis. He has more than 50 publications in reputed journals, reports, and conferences.

Sanjay Chatterji has received Ph.D. from IIT Kharagpur on Indian Language Machine Translation domain specifically on statistical and transfer-based approaches to build Bengali-Hindi machine translation systems based on the resources that are available. He has built morphological analyzer, POS tagger, parser, etc., for Bengali language. In Samsung R&D Institute India Bangalore (SRIB), he has worked in the domain of semantic analysis, specifically on aspect-based sentiment analysis, inter concept higher order predicate relation extraction, representation and application in search domain, DNN-based classification of intents, etc. Overall, he has experience and expertise in application of machine learning algorithms for NLP tasks. He has worked in SRIB for about 4 years and in multiple engineering colleges as Lecturer and Assistant Professor for 3 years. He has published 4 journal papers (3 of which are SCI indexed) and 16 conference papers and achieved jury award for a demo in Nipun, 2015, in SRIB. Currently, he is working as an Assistant Professor in Indian Institute of Information Technology Kalyani, West Bengal, India.

Pankaj K. Sa received the Ph.D. degree in Computer Science in 2010. He is currently serving as an Associate Professor with the Department of Computer Science and Engineering, National Institute of Technology Rourkela, India. His research interest includes computer vision, biometrics, visual surveillance, and robotic perception. He has co-authored a number of research articles in various journals, conferences, and book chapters. He has co-investigated some research and development projects that are funded by SERB, DRDO-PXE, DeitY, and ISRO. He is the recipient of prestigious awards and honors for his excellence in academics and research. Apart from research and teaching, he conceptualizes and engineers the process of institutional automation.

Applications of Neural Network

Minutiae Points Extraction Using Faster R-CNN

Vivek Singh Baghel, Akhilesh Mohan Srivastava, Surya Prakash, and Siddharth Singh

Abstract A fingerprint is an impression of the friction ridges of all parts of the finger and it is a widely used biometric trait. In a fingerprint authentication system, minutiae points are used as features to authenticate an individual. There are many traditional techniques which have been proposed to extract minutiae points from a fingerprint. However, efficiently locating minutiae points in a fingerprint image is still a challenging problem. In this paper, we propose a technique to locate and classify the minutiae points based on Faster R-CNN, which is a combination of region proposal network (RPN) and detection network. We use IIT Kanpur fingerprint database to evaluate the proposed technique and to show the performance.

Keywords Biometrics · Fingerprint · Minutiae Extraction · CNN · Faster R-CNN

V. S. Baghel (✉) · A. M. Srivastava · S. Prakash
Indian Institute of Technology Indore, Indore 453552, India
e-mail: phd1801201005@iiti.ac.in

A. M. Srivastava
e-mail: phd1701101001@iiti.ac.in

S. Prakash
e-mail: surya@iiti.ac.in

S. Singh
National Institute of Technology Hamirpur, Hamirpur 176005, India
e-mail: ece16447@nith.ac.in

J. K. Mandal et al. (eds.), *Computational Intelligence and Machine Learning*, Advances in Intelligent Systems and Computing 1276,
https://doi.org/10.1007/978-981-15-8610-1_1

1 Introduction

A fingerprint is a combination of continuous patterns which are known as ridges and valleys. In each fingerprint, some unique points exist which are called minutiae points and these points are used to discriminate a fingerprint from others. Bifurcation and ridge-end are mainly used minutiae points in a fingerprint authentication system. The locations where ridges bifurcate into two different ridges are called bifurcation points, whereas points, where ridges end abruptly, are called ridge-end points. For the last several years, deep convolutional neural networks (CNNs) [4] has been used to solve several object recognition problems. The basic architecture of a CNN consists of several building blocks, including different layers such as convolution layers, pooling layers (e.g., max pooling and min pooling), fully connected layers, and classification layer. CNNs have also been used for extracting the minutiae points from a fingerprint image [3]. R-CNN, which is also called as regions with CNN features, has been widely used to detect the locations of objects present in a scene with their class labels. The extraction of the location of minutiae points with their types (bifurcation or ridge-end) can also be mapped to object detection problem in an image containing multiple objects. Although R-CNN achieves great results, the training process has lots of problems because the region proposal has to be generated for the training dataset. In R-CNN, region proposals are calculated by the selective search method, which is a time-consuming process. The improved version of R-CNN is Fast R-CNN, which uses CNN completely for the whole image and RoI pooling is performed on the feature map for classification and regression. However, it still uses a selective search method. The improved version of Fast R-CNN is called Faster R-CNN. It is a fully convolutional network and is fast in training process as compared to both R-CNN and Fast R-CNN. In this network, RPN is used to propose the regions where the object of interest may be present in a scene. This motivates us to use Faster R-CNN for the extraction of minutiae points in a fingerprint image. To carry out the same, training data has to be prepared, which contains fingerprint images with ground truth bounding boxes representing the minutiae locations and their respective class labels (i.e., bifurcation or ridge-end).

Rest of the paper is organized as follows. Section 2 discusses some of the existing techniques used for minutiae points extraction. Description of Faster R-CNN is given in Sect. 3. The proposed method is discussed in the Sect. 4. Experimentation and results are discussed in the Sect. 5. The last section discusses the conclusions.

2 Literature Review

Fingerprint biometrics is a widely explored research area in biometrics and there are many works that have been carried out in this area. Ratha et al. [13] proposed a method to extract the minutiae points based on orientation flow and they have compared their result with the ground truth (the manually extracted minutiae points). Jain et al. [6] have proposed an online fingerprint verification system and their minutiae extraction algorithm is shown to be much faster and reliable as compared to the technique proposed in [13]. Hong et al. [11] have proposed a method of enhancing the fingerprint image to extract minutiae points correctly. Jain et al. [8] proposed a new representation of features in a fingerprint image instead of minutiae points. This feature representation is called fingercodes. Jain et al. have explained in [7, 9] about the importance of fingerprint authentication. They have also explained the feature extraction of a fingerprint image and the matching techniques for fingerprint authentication. The features, which are basically minutiae points, are extracted by traversing a 3×3 filter onto the thinned fingerprint image. Lavanya et al. [2] proposed a Gabor filter-based enhancement algorithm for minutiae extraction, which can simultaneously extract ridge orientation and ridge frequency. Other than above-mentioned techniques, there are many works which have been carried out in the area of fingerprint authentication for minutiae points extraction based on traditional approaches.

In recent years, studies show that deep neural network architectures [4] are potential candidates for feature extraction. In the domain of minutiae extraction, there are many works which use deep neural networks for minutiae extraction. Jiang and Liu [10] presents fingerprint minutiae detection based on multi-scale CNN and employ two separate CNN for minutiae detection and minutiae classification. In their work, the minutiae detection is carried out using two steps. In the first step, classification of image patches into minutiae and non-minutiae patch is carried out, whereas, in the second step, classification is done for minutiae into ridge-end and bifurcation. Tang et al. [15] have proposed an approach to extract minutiae points from latent fingerprint images and have compared the accuracy of the authentication system with the accuracy, while minutiae points are calculated manually. Jiang et al. [12] proposed a novel minutiae extraction approach by using two nets, i.e., judge net and locate net. In this work, fingerprint image has been cropped into several overlapped patches. Classification of these patches into minutiae and non-minutiae is carried out using judge net, whereas locate net in this work is used to locate the minutiae points into the overlapped patches.

3 Proposed Method

We propose an automated approach for the extraction of minutiae points from a fingerprint image. The technique is based on Faster R-CNN [14] which is a fully convolutional neural network and is end-to-end trainable. The complete schematic of Faster R-CNN used in the proposed technique for extraction of minutiae points is shown in Fig. 1. As discussed in the introduction section, there are different types of minutiae points present in a fingerprint image. Here, bifurcation and ridge-end points are used as the minutiae points and are detected in a fingerprint image.

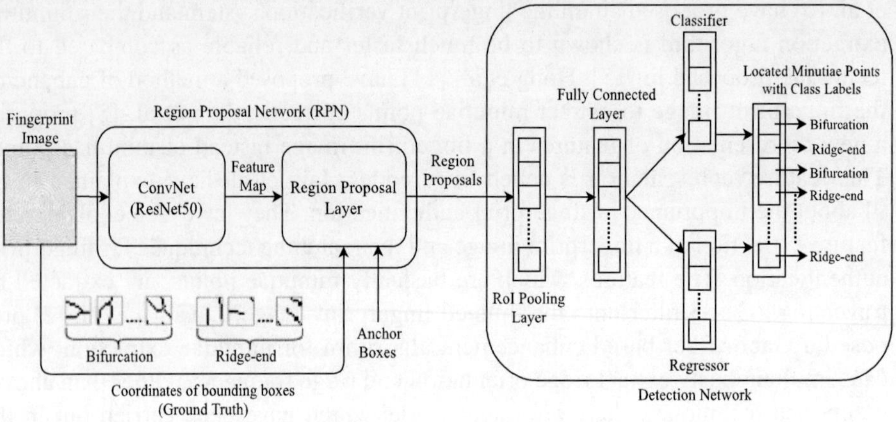

Fig. 1 The schematic diagram of Faster R-CNN used in fingerprint minutiae extraction

3.1 Ground Truth Preparation

The preparation of ground truth is an essential step before performing training of the network in the proposed technique. Ground truth for training the network are patches of size $m \times m$ around the minutiae points present in a fingerprint image. These are represented in the form of bounding boxes and the coordinates of bounding boxes are used for training purpose. Coordinates of these bounding boxes along with class label (minutiae type, viz. bifurcation or ridge-end) are stored in an annotation file. At the time of training of Faster R-CNN, this annotation file is used as an input to provide the details of the ground truth.

3.2 Training of Faster R-CNN

Faster R-CNN is mainly a combination of RPN and detection network. RPN is a combination of ConvNet, region proposal layer, and two parallel layers classification and regression. Pre-trained ResNet50 [5] is used as a ConvNet in our proposed approach and hence there is no training required for ResNet50. ResNet50 is a convolutional neural network which is used to classify the objects. It generates feature maps from the fingerprint images and these feature maps are the output of the last convolution layer or the base layers of ResNet50. These feature maps and ground truth information obtained in the previous step are further feed-forwarded to the region proposal layer. Region proposal layer uses anchor boxes of different sizes and sliding-window strategy to find out the maximum overlap of anchor boxes with ground truth. In RPN, after the region proposal layer, two parallel layers which are classification and regression layers exist. Both of these layers are also convolution layers. The training (fine-tuning) of RPN is performed using pre-trained weights of base layers, anchor boxes of different sizes, fingerprint images, and the ground truth. The tuning of weights is mainly performed for region proposal, classification, and regression layers of RPN, whereas its other remaining part uses pre-trained weights of ResNet50. The region proposals generated by RPN are used to train the detection network. RoI pooling layer, fully connected layers, and two parallel layers, i.e., classification and regression layers are the part of the detection network. Classification and regression layers are the fully connected layer in the detection network. The training of detection network is performed using region proposals generated by the RPN. Backpropagation is used to train both RPN and detection network.

3.3 Localization of Minutiae Points

After training the Faster R-CNN, to carrying out localization of minutiae points, a fingerprint image is given to the network as an input. In RPN, ConvNet generates the feature map which is further feed-forwarded to the region proposal layer. Region proposal layer computes the region proposals based on the anchor box sizes which have been used during the training of RPN. These region proposals are further feed-forwarded to the detection network where RoI pooling layer flattens these regions in the form of feature vectors of same sizes. Computed feature vectors are further used to classify and detect the location of minutiae points by fully connected, classification, and regression layers of the detection network. At the end, location and type of minutiae points are obtained as outputs.

4 Experimental Analysis

This section evaluates the performance of the proposed technique. It also provides the details of the database and the values of hyperparameters. We have used IIT Kanpur fingerprint database for carrying out the experimentation. This database contains 1376 subjects where for each subject, there are four samples present in the database. This gives a total of 5504 images for training, validation, and testing.

The preparation of ground truth is a very important step to train the network on a given dataset. To accomplish the same, first minutiae points are extracted by using VeriFinger Demo Version [1]. Further, the bounding boxes of size 13×13 are calculated around the extracted minutiae points for each fingerprint image. Further, the training is performed on thinned fingerprint images. For thinning of fingerprint images, there are mainly two steps involved. The first step performs the enhancement on fingerprint images by using Gabor filters of different orientations and convert images into binary. In the second step, by using a morphological function, enhanced binary images are converted into thinned binary images. These thinned images are used for training purposes. During the training, Faster R-CNN takes an annotation file as an input. Training has been performed on 70% images of the total fingerprint

Table 1 Hyperparameter values for Faster R-CNN used in the proposed technique

Hyperparameter	Value
Optimizer-RPN and Detection network	"Adam"
Learning rate	0.00001
Number of epochs	20
Epoch length	128
Anchor box sizes	128, 256, and 512 px
RPN Stride	16
Overlaps for classifier RoIs	Minimum-0.1, Maximum-0.5
Overlaps for RPN	Minimum-0.3, Maximum-0.7
Non-maximum suppression	Overlap threshold-0.8

images available in the database. Remaining 30% images are used to validate and test the proposed technique. The values of hyperparameters for the training of Faster R-CNN are given in Table 1. It contains different parameter values like types of optimizer, learning rate, number of epoch, epoch length, anchor boxes scale, etc. Anchors also play a very important role in the training of the Faster R-CNN network. They are actually bounding boxes of different sizes that are used to propose the region by RPN. The ratios between width and height of boxes are chosen as 0.5, 1.0, and 1.5. All possible combinations of these ratios and the size of anchors are used to perform the training of the network.

The results of the proposed technique on some sample images of the database have been shown in Fig. 2 where ground truth and detected minutiae points are shown. As we can see, the proposed technique detects minutiae points quite efficiently. In Fig. 2a, it can be clearly seen that most of the minutiae points are extracted when we compare it with the ground truth. In some cases, the number of detected minutiae points using Faster R-CNN is less as compared to the ground truth. An example of such cases is depicted in Fig. 2b. It is evident from Fig. 2 that even though the less number of minutiae points are detected in some cases, their locations and the class labels (either bifurcation or ridge-end) are predicted correctly. This shows that the precision value of the proposed technique is high. These results are significant to show that the proposed technique has worked quite well and this can be further improved in future to extract minutiae points in more challenging fingerprint images such as noisy, wet, and dry fingerprint images.

5 Conclusions

In this work, a new technique has been proposed to extract minutiae points from a fingerprint image using Faster R-CNN. Faster R-CNN, which is a combination of RPN, and detection network, has a good capability in locating specific objects

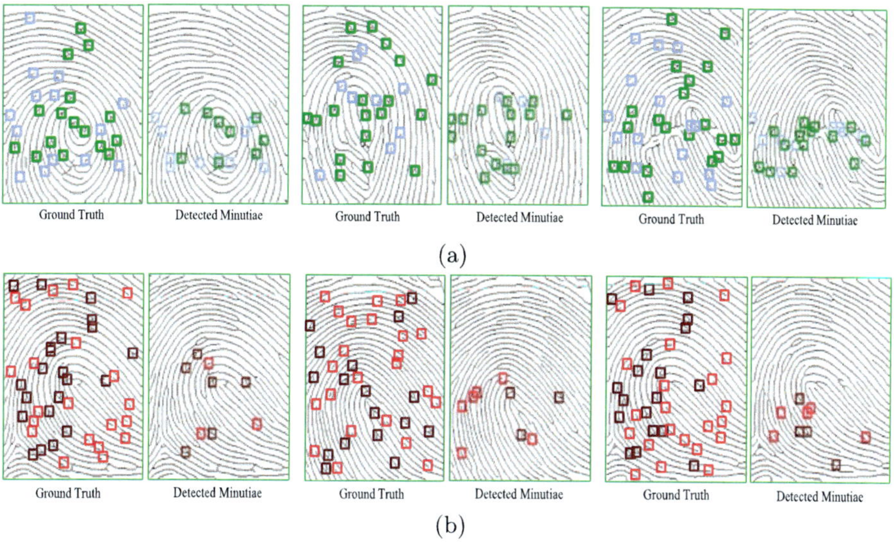

Fig. 2 Results of minutiae points detection using the proposed technique: **a** Sample images with large number of detected minutiae points, **b** Sample images with relatively lesser number of detected minutiae points as compared to (**a**)

in a scene. We exploit it for the purpose of locating minutiae points in a fingerprint image where the specific objects (patches) being located are the regions having ridge-end and the bifurcation points at its center in a fingerprint image. In the proposed technique, the first feature map is computed from a fingerprint image by using ConvNet which is a part of RPN and further, is feed-forwarded to region proposal layer. Region proposal layer takes feature map and ground truth as inputs and generates region proposals which are subsequently feed-forwarded to detection network. RoI pooling layer which is a part of detection network pools the region proposals into feature vectors. By using these feature vectors, the types of minutiae points and the bounding boxes around their locations are computed by the classifier and regressor of the detection network. The proposed technique has been evaluated on IIT Kanpur fingerprint database. The detected minutiae points in a fingerprint image are compared with the ground truth. It has been observed that the proposed technique shows encouraging performance in the detection of minutiae points.

References

1. Goodfellow, I., Bengio, Y., Courville, A.: Deep Learning. The MIT Press (2016)
2. Cao, K., Jain, A.K.: Automated latent fingerprint recognition. IEEE Trans. Pattern Anal. Mach. Intell. **41**(4), 788–800 (2019)
3. Ratha, N.K., Chen, S., Jain, A.K.: Adaptive flow orientation-based feature extraction in fingerprint images. Pattern Recogn. **28**(11), 1657–1672 (1995)

4. Jain, A., Lin Hong, Bolle, R.: On-line fingerprint verification. IEEE Trans. Pattern Anal. Machine Intell. **19**(4), 302–314 (1997)
5. Hong, L., Wan, Y., Jain, A.: Fingerprint image enhancement: algorithm and performance evaluation. IEEE Trans. Pattern Anal. Mach. Intell. **20**(8), 777–789 (1998)
6. Jain, A.K., Prabhakar, S., Hong, L., Pankanti, S.: Filterbank-based fingerprint matching. IEEE Trans. Image Process. **9**(5), 846–859 (2000)
7. Jain, A.K., Flynn, P., Ross, A.A.: Handbook of Biometrics, 1st edn. Springer Publishing Company, Incorporated (2010)
8. Jain, A.K., Ross, A.A., Nandakumar, K.: Introduction to Biometrics. Springer Publishing Company, Incorporated (2011)
9. Lavanya, B.N., Raja, K.B., Venugopal, K.R., Patnaik, L.M.: Minutiae extraction in fingerprint using gabor filter enhancement. In: Proceedings of International Conference on Advances in Computing, Control, and Telecommunication Technologies, pp. 54–56 (2009)
10. Jiang, H., Liu, M.: Fingerprint minutiae detection based on multi-scale convolution neural networks. In: Proceedings of Chinese Conference of Biometric Recognition, vol. 10568, pp. 306–313 (2017)
11. Tang, Y., Gao, F., Feng, J.: Latent fingerprint minutia extraction using fully convolutional network. In: Proceedings of IEEE International Joint Conference on Biometrics, pp. 117–123 (2017)
12. Jiang, L., Zhao, T., Bai, C., Yong, A., Wu, M.: A direct fingerprint minutiae extraction approach based on convolutional neural networks. In: Proceedings of International Joint Conference on Neural Networks, pp. 571–578 (2016)
13. Ren, S., He, K., Girshick, R., Sun, J.: Faster R-CNN: Towards real-time object detection with region proposal networks. In: Proceedings of Neural Information Processing Systems, pp. 91–99 (2015)
14. He, K., Zhang, X., Ren, S., Sun, J.: Deep residual learning for image recognition. In: Proceedings of IEEE Conference on Computer Vision and Pattern Recognition, pp. 770–778 (2016)
15. Verifinger SDK Demo. https://www.neurotechnology.com/verifinger.html. [Online; Accessed 26 Oct 2019]

Genetic Algorithm-Based Optimization of Clustering Data Points by Propagating Probabilities

Shailja Dalmia, Aditya Sriram, and T. S. Ashwin

Abstract Clustering is among the pivotal elementary operations in the field of data analysis. The efficiency of a clustering algorithm depends on a variety of factors like initialization of cluster centers, shape of clusters, density of the dataset, and complexity of the clustering mechanism. Previous work in clustering has managed to achieve great results but at the expense of a trial and error approach to achieve optimal values for user-defined parameters which have a huge bearing on the quality of the clusters formed. In this work, we propose a solution that optimizes the user-defined parameters for clustering algorithm called Probability Propagation (PP) by harnessing the capabilities of Genetic Algorithm (GA). In order to overcome this sensitivity in PP, a novel optimization technique is applied by obtaining the optimal values of δ and s using GA by maximizing inter-cluster spread and minimizing intra-cluster spread among the clusters being formed. The proposed method was found to retrieve top chromosomes (bandwidth and s) with a similar number of clusters, thus eliminating the sensitivity of user-defined parameters which is optimized by using GA.

Keywords Data Clustering · Probability Propagation · Density · Genetic Algorithm

S. Dalmia · A. Sriram · T. S. Ashwin (✉)
National Institute of Technology Karnataka, Surathkal Mangalore 575025, Karnataka, India
e-mail: ashwindixit9@gmail.com

S. Dalmia
e-mail: shailjadalmia11@gmail.com

A. Sriram
e-mail: adityasriram1995@gmail.com

J. K. Mandal et al. (eds.), *Computational Intelligence and Machine Learning*, Advances in Intelligent Systems and Computing 1276,
https://doi.org/10.1007/978-981-15-8610-1_2

1 Introduction

Over the past few decades, applications of clustering have tremendously grown in varied fields as a precursor for data science use cases [4]. The classical k-means clustering algorithm suffers from preliminary centers for clustering which has an adverse effect on the quality of clusters and the speed of convergence. To solve the issue of arbitrary initialization, Frey and Dueck [1] had put forth an effective method for clustering called Affinity Propagation (AP). The flipside to this solution is the complex nature of message passing instructions employed by this method.

Motivated by AP, the PP can pin-point clusters with both spherical and non-spherical shapes. It starts with a stochastic matrix of probabilities and propagates iteratively until a group of attractors become stable. The AP, PP, and Markov Clustering (MCL) use message passing techniques but the rules of message passing is much simpler in case of PP with the initialization of stochastic matrix being much different in this case as compared to the other two. The quality of most clustering algorithms is deeply influenced by the initialization parameters which result in different clustering results for varying parameters.

To this end, the PP algorithm does suffer from initialization of two user-defined parameters, namely bandwidth and s which controls the neighbors set and shape of clusters, respectively, and thus forms clusters based on the initialization of user-defined parameters.

In order to address this drawback in PP algorithm, a novel optimization technique based on Genetic Algorithm (GA) has been proposed.

2 Related Work

Even though clustering plays a pivotal role in the modern applications, very few algorithms are in place for optimization of such clustering algorithms to retrieve better clustering results which are insusceptible to any user-defined parameters. To this end, Frey et al. [1] proposed AP clustering which does not necessitate the amount of clusters to be recognized beforehand and has edge over other algorithms based on efficiency and precision but is not appropriate for large-scale data clustering. Dongen et al. [2] in their work were able to come up with MCL algorithm which is a simulation of random flow in the graph, however, the overall process is very slow. DBSCAN is a density-based clustering algorithm which classifies points in space as core, non-core, and outliers points where high-density region cluster together with neighboring points leaving outliers to lie in low-density region as different clusters are formed, but appropriate combination of E radius and minPts cannot be determined in case of large datasets [3].

Each of the existing clustering algorithms has their own shortcomings based on different parameters. We aim to conceive optimization for PP clustering algorithm which identify non-spherical and spherical clusters with varying density by employing a message passing technique with optimal values of initialization parameters using GA such that it can perform well on large as well as small datasets.

3 Methodology

The proposed optimization of PP algorithm is performed using GA to eliminate the need for user-defined parameters like bandwidth and s and provide consistent clustering results. The work by Gana et al. [4] was incorporated along with the optimization of parameters to minimize intracluster distance and maximize intercluster distance between data points.

Figure 1 represents the flow diagram of the proposed algorithm. The idea is to evolve the parameter by selecting the fittest chromosomes which would result in better clustering results rather than varying parameters which would differ the results.

3.1 *Optimized PP Algorithm with Genetic Algorithm*

The PP algorithm [4] deals with two main user-defined parameters which governed the cluster orientation in the space. The bandwidth parameter was used as a threshold value between two data point. The other parameter called s determines the cluster size. In our work, the proposed model is to optimize these user-defined parameters by applying GAs [5] to find the ideal values of bandwidth and s such that an optimal number of clusters can be found. GA begins with the initialization of population with the selection of the fittest chromosomes accompanied by crossover and mutation.

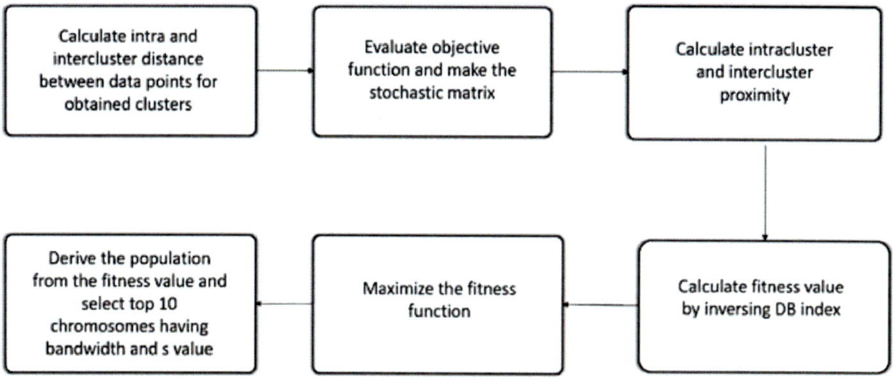

Fig. 1 Flow diagram of GA-incorporated PP algorithm

There are two main steps in the procedure of the application of GA:

1. **Individual Representation** In the proposed procedure, the chromosomes consists of randomized real numbers to symbolize bandwidth and s. The length of chromosomes is 2 (bandwidth and s). Initially, the population size is fixed and random initialization of k (population size) chromosomes is done with 2*k genes values of bandwidth and s.
2. **Evaluation of Fitness** The purpose of clustering is to rift data into clusters. A resultant barrier should have below given assets:
   - Homogeneity: Data going to the identical cluster should be as alike as probable.
   - Heterogeneity: Data going to the dissimilar cluster should be as unlike as probable.
3. The distinction of the data points to be in same or different clusters (as pointed above) is done on the basis of the distance between them.

In the proposed method, the Davies–Bouldin (DB) index is used to evaluate the ideality of the clusters which helps to calculate the fitness assessment of the discrete chromosome. The techniques used below have been used in [13] to identify the best number of clusters, however, in this work; a novel approach is proposed to optimize the user-defined constraints through these techniques as will be explained subsequently. The DB index is applied in [9, 10, 12] also to gauge the soundness of the clusters. This is defined as the ratio of the sum of cluster scatter which is within the separation between clusters described in [13]. Then the DB index is demarcated as follows:

$$DB = \frac{1}{K}\sum_{i=1}^{K} R_{i,qt}. \tag{1}$$

Algorithm 1 Optimized PP Algorithm with GA

```
Randomize initialization of bandwidth parameter and s
Initialize the number_of_generations(NGEN), population_size, mutation_rate, and crossover_rate.
while NGEN >0 do
    for each bandwidth and s do
        Obtain the clusters as specified in Algorithm 1.
        Calculate the intercluster and intracluster distances [13]
        Calculate the DB index as per Eq. (1)
        Calculate fitness value as 1/DB and maximize it.
    end for
    Apply crossover and mutation operator over the chromosomes.
    NGEN–
end while
select the top 10 fittest chromosomes
```

A small value of DB index would be most fit, hence to minimize DB index $1/DB_i$ is maximized (DB index calculated for discrete i). Thus, maximum fitness value chromosome is selected to perform crossover and mutation [6–8]. Crossover is two-point crossover and a fixed crossover rate of 50% is applied to be performed on highly fit chromosome.

To incorporate variety, mutation is performed after crossover for a fixed mutation rate of 10%. After performing one iteration, the above procedure is recurred for a static number of iterations after which highly fit chromosomes denote the user-defined bandwidth and s upon which the PP algorithm is used to get an optimal number of clusters. Algorithm 2 presents the optimized PP algorithm through GA. Lines 1 and 2 deal with the initialization of various parameters. Lines 3–11 iterate over the number of generations, and evaluate the fitness value for every chromosome. Line 10 applies the biological operators like crossover and mutation over the chromosomes to incorporate variety and produce another set of chromosomes for the next generation. Line 13 selects the top 10 fittest chromosomes, giving us the optimal value of bandwidth and s.

3.2 *Result and Analysis*

This section deals with the detailed analysis of the PP algorithm along with GA for a favorable number of clusters for the two datasets; the synthetic dataset which consists of 40 data points with pre-defined class labels; and the other one is a real dataset of 62 data points which is Alizadeh V2 dataset of genetic representation of lung cancer cells. Results were recorded for triangle and uniform kernel functions for both synthetic and real datasets. The results for the real dataset for triangular kernel consist of the number of clusters we get for each of s = 1, 7, and 62. The user-defined parameters here are the bandwidth parameter and s. The analysis shows that the number of clusters we get for each of the s parameters are approximately the same. Tables 1, 2, 3 and 4 summarize the results for the same.

The PP algorithm gives the number of clusters for given values of bandwidth parameters and s values; now GA is applied for getting the best optimal values

Table 1 The time (T), number of clusters (K), and fitness values (FV) are recorded for given parameters when applied to a synthetic dataset of 40 data points for Uniform kernel

δ	K	T	FV	δ	K	T	FV	δ	K	T	FV
s = 1				s = 18				s = 27			
6.3	25	0.285	0.066	6.3	25	0.265	0.066	6.3	25	0.281	0.0666
11.61	9	0.602	0.226	11.61	8	1.081	0.260	11.61	8	1.284	0.026
16.15	6	0.584	0.418	16.15	5	1.276	0.641	16.15	5	1.405	0.6425
30.38	3	0.439	0.426	30.38	2	1.274	0.019	30.38	2	1.235	0.019

Table 2 The time (T), number of clusters (K), and fitness values are recorded for given parameters when applied to a synthetic dataset of 40 data points for Triangle kernel

δ	K	T	FV	δ	K	T	FV	δ	K	T	FV
s = 1				s = 18				s = 27			
6.3	25	0.880	0.0666	6.3	25	0.757	0.0666	6.3	25	0.774	0.0666
11.61	11	0.681	0.1592	11.61	8	1.067	0.26	11.61	8	1.092	0.26
16.15	8	0.556	0.2479	16.15	5	1.304	0.6415	16.15	5	1.257	0.6415
30.38	3	0.423	0.8410	30.38	3	0.776	0.8410	30.38	3	0.982	0.8410

Table 3 The time (T), number of clusters (K), and fitness values (FV) are recorded for given parameters when applied to Alizadeh-V2 dataset for Uniform kernel

δ	K	T	FV	δ	K	T	FV	δ	K	T	FV
s = 1				s = 7				s = 62			
0.4	49	1.47	0.0225	0.4	50	1.47	0.0224	0.4	50	1.48	0.0214
0.59	29	1.43	0.0478	0.59	29	1.49	0.0478	0.59	29	1.57	0.0478
0.67	11	1.96	0.7857	0.67	10	3.19	0.5882	0.67	10	3.31	0.5882
0.79	3	1.99	0.75	0.79	3	2.02	0.6	0.79	3	2.21	0.6

Table 4 The time (T), number of clusters (K), and fitness values (FV) are recorded for given parameters when applied to Alizadeh-V2 dataset for Triangle kernel

δ	K	T	FV	δ	**K**	T	FV	δ	K	T	FV
s = 1				s = 7				s = 62			
0.4	49	1.44	0.0224	0.4	49	2.54	0.0224	0.4	49	2.56	0.0224
0.5	29	1.44	0.0485	0.59	29	2	0.0485	0.59	29	2	0.0485
0.67	12	2	0.6	0.67	10	4.26	0.6	0.67	10	4.31	0.6
0.79	3	1.43	0.75	0.79	3	2	0.75	0.79	3	2.23	0.75

Table 5 The top chromosomes for each population set are recorded which consist of bandwidth parameter and s values. The number of clusters (K) derived from them are also recorded. The results are recorded for 5, 10, and 15 number of generations with a mutation rate of 10% and crossover rate of 50%

Number of Generations = 5			Number of Generations = 10			Number of Generations = 15		
δ	s	K	δ	s	K	δ	s	K
0.856	2	36	1.03	3	35	0.156	2	33
0.157	3	38	0.12	3	36	0.488	3	35
0.157	3	40	1.26	3	35	0.488	3	36
1.273	3	39	0.129	3	34	0.488	3	35
1.273	3	40	0.129	3	36	0.156	2	33

for bandwidth and s. The results for the same have been summarized in Table 5. The results are recorded for the top five chromosomes for 5, 10, and 15 number of generations. The population size has been fixed to ten chromosomes, with each chromosome consisting of two genes—one for bandwidth parameter and other for s value. The number of clusters (K) related to these values is also recorded which give us the most favorable number of clusters for the derived s and bandwidth parameter through GA.

4 Conclusion and Future Work

The PP algorithm performs better than other classical clustering algorithms like AP. When compared to the time taken, PP algorithm performs at par with other algorithms. PP algorithm does well for synthetic as well as for a sparse real dataset for any shape of the cluster. As an improvement over the PP algorithm, a novel approach is formulated to calculate the optimal number of clusters by harnessing the capabilities of the GA. At the core of this approach, we formulate the fitness function such that the inter-cluster distance between data points is reduced and intra-cluster distance in between clusters is increased. The fitness value of each cluster model is recorded and a maximizing fitness function is applied to it to record the best bandwidth parameter and ‘s’ value for which an optimal number of clusters is obtained. For further optimizations and comparisons, different genetic operators can be explored other than mutation and crossover. Parallelizing techniques for reducing the time complexity of the proposed GA-based PP algorithm can be incorporated, which is currently $O(n^3)$.

References

1. Gana, G., Zhangb, Y., Dey, D.K.: Clustering by propagating probabilities between data points. Appl. Soft Comput. **41**, (2016)
2. Frey, B.J., Dueck, D.: Clustering by passing messages between data points. Science **315** (2007). www.sciencemag.org
3. van Dongen, S.: Graph clustering via a discrete uncoupling process? SIAM J. Matrix Anal. Appl. **30**(1), 121–141 (2008)
4. Ester, M., Kriegel, H.-P., Sander, J., Xu, X.: A density-based algorithm for discovering clusters in large spatial databases with noise. KDD (1996)
5. Goldberg, D.E.: Genetic Algorithms in Search, Optimization and Machine Learning. Addison-Wesley (1989)
6. Liu, Y., Ye, M., Peng, J., Wu, H.: Finding the optimal number of clusters using genetic algorithms. In: IEEE Conference on Cybernetics and Intelligent Systems. Chengdu, pp. 1325–1330 (2008)
7. Bandyopadhyay, S., Maulik, U.: Nonparametric genetic clustering: comparison of validity indices. IEEE Trans. Syst. Man Cybern. Part C: Appl. Rev. **31**(1), 120–125 (2001)
8. Bandyopadhyay, S., Maulik, U.: Genetic clustering for automatic evolution of clusters and application to image classification. Pattern Recogn. **35**(6), 1197–1208 (2002)

9. Lai, C.C.: A novel clustering approach using hierarchical genetic algorithms. Intell. Autom. Soft Comput. **11**(3), 143–153 (2005)
10. Kumsawat, P., Attakitmongcol, K., Srikaew, A.: A new approach for optimization in image watermarking by using genetic algorithms. IEEE Trans. Signal Process. **53**(12), 4707–4719 (2005)
11. Ramasubramanian, P., Kannan, A.: A genetic-algorithm based neural network short-term forecasting framework for database intrusion prediction system. Soft Comput. **10**(8), 699–714 (2006)
12. Chang, Y.C., Chen, S.M.: A new query reweighting method for document retrieval based on genetic algorithms. IEEE Trans. Evol. Comput. **10**(5), 617–622 (2006)
13. Lin, H.J., Yang, F.W., Kao, Y.T.: An efficient GA-based clustering technique. Tamkang J. Sci. Eng. **8**(2), 113–122 (2005)

Detection of Malaria Parasites in Thin Blood Smears Using CNN-Based Approach

Sabyasachi Mukherjee, Srinjoy Chatterjee, Oishila Bandyopadhyay, and Arindam Biswas

Abstract Early and accurate detection of malaria is the main key in controlling and eradicating this deadly disease. An efficient automated tool for examining stained blood slides can be very helpful in this regard. This paper proposes an approach for automatic detection of malarial parasites from thin blood smears. Texture- and intensity-based analyses are performed to detect the blood particles in the pre-processed image. Extracted red blood cell particles are then sent to a convolutional neural network for further probing. The proposed CNN architecture is trained with a publicly available dataset along with some Giemsa-stained blood smear images collected from a hospital. The performance of malaria detection process gives a satisfactory dice score of 0.95.

Keywords Malaria detection · CNN · RBC segmentation · Thin blood smear

1 Introduction

Malaria is a worldwide threat that causes numerous deaths every year. Protozoan parasites are responsible for malaria which are transmitted by the female Anopheles mosquitoes. According to the World malaria report 2018, there were 219 million

S. Mukherjee (✉) · A. Biswas
IIEST Shibpur, Howrah, India
e-mail: sabya84@gmail.com

A. Biswas
e-mail: barindam@gmail.com

S. Chatterjee
VIT University, Vellore, India
e-mail: srinjoychatterjee27@gmail.com

O. Bandyopadhyay
IIIT Kalyani, Nadia, India
e-mail: oishila@gmail.com

J. K. Mandal et al. (eds.), *Computational Intelligence and Machine Learning*, Advances in Intelligent Systems and Computing 1276,
https://doi.org/10.1007/978-981-15-8610-1_3

cases of the disease in 2017 which is very alarming. The main reason behind so high mortality rate is the delay in disease detection as well as unavailability of life-saving infrastructure. To defeat malaria completely, a combined approach of mosquito control measures along with early diagnosis and treatment are needed. Transmission of the malaria parasite (plasmodium) takes place when anopheles feeds on human blood. The main parasites that are responsible for malaria include P.falciparum, P.malariae, P.vivax and P.ovlae. The infected blood cells change the appearance by which malaria can be tracked by looking at the stained blood smear [7]. Four main erythrocytic stages of malaria parasite affects the red blood cells (RBC), namely ring, trophozoite, schizont and gametocyte. The changes caused by these stages to the healthy RBC can be viewed by staining the diseased blood smear using Giemsa method [12]. Depending on severity and stage of malaria, the appearance of RBCs differ, but the main noticeable changes are the presence of parasite cytoplasm and the colour difference of the affected cells from that of the healthy cells. As the goal is to detect the presence of malaria, any discrepancies between expected texture and colour inside RBC indicates the presence of parasite. As can be shown from Fig. 1a, there is no contaminated RBC, while in Fig. 1b, malaria-affected RBCs can be seen clearly.

The most challenging part of malaria detection is the minute study of the blood smears by human experts. The process is very time-consuming and heavily dependent upon the efficiency of the observer. It results in delay in disease diagnosis. There is always a chance of human error in such observations. Moreover, in rural areas, the availability of trained human resource for blood smear examination is very limited. An automated process will reduce both diagnosis time and chance of human error. This paper proposes a method that performs automated detection of malaria from blood images of thin blood smears by using convolutional neural networks (CNN). Before passing images to CNN, some pre-processing activities are performed to achieve better accuracy and faster computation time.

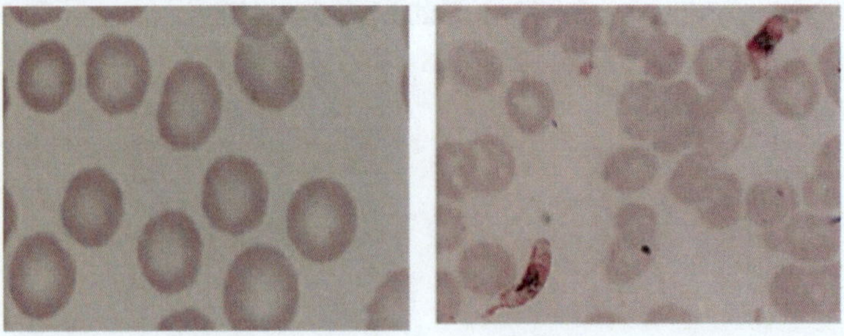

(a) healthy blood smear (b) Malaria affected blood smear

Fig. 1 Images of thin blood smear

2 Related works

Researchers are working towards automated detection of malaria for the past few years. The main focus of malaria detection process is to separate the RBCs and classify the parasites by applying the conventional image processing and machine learning approaches. Somasekar et al. [15] proposed an efficient algorithm for contrast enhancement of the Giemsa-stained blood smears images using γ order image a look-up-table. Khan et al. [3] have made an effort for parasites using the unsupervised k-means clustering using green channel of the original method. KNN classifier-based identification of life cycle stages of malaria parasites has been proposed by Nanoti et al. [8]. Reni et al. [13] suggested pre-processing of the image using morphological operations. Plasmodium falciparum and plasmodium vivax parasite identification is done by Penas et al. [11] using CNN architecture inception v3-based approach. Mashor et al. [6] have applied K-means clustering-based analysis on Giemsa-stained blood smear images to identify the presence of malaria parasites. A review of the existing technologies with a detail discussion on segmentation of malaria parasite along with feature extraction and classification of the parasite have been done in [12]. Object detection in thin blood smear by a three-stage method and Kalman filter has been proposed by [10]. Malaria cell segmentation using HSI and CIE colour space followed by a region-growing approach is done by [16]. Kaur et al. [2] suggested a way of detecting edges of the malaria parasites present in blood smear images using ant colony optimization algorithm.

3 Proposed Method

The proposed method consists of two main components. The first part is to pre-process and segment the suspected RBC and the second part is to classify it as healthy or parasitized. The complete flow of the process is shown in Fig. 2. The process starts with intensity enhancement that helps in the subsequent phases. Segmentation of RBCs is done using entropy and intensity profile-based analysis of the images. These cells (RBC) are then extracted from the main image and resized. The processed individual RBC images are sent to CNN for detecting the presence of malaria parasites.

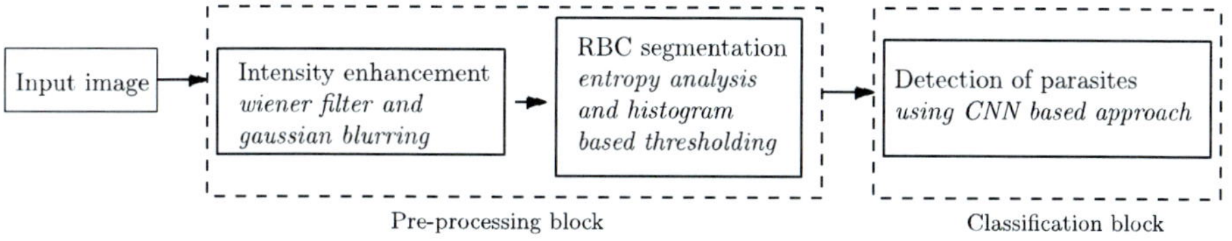

Fig. 2 Block diagram of the proposed approach

3.1 Intensity Enhancement

Intensity enhancement and noise removal are important parts of the pre-processing phase as the performance of successive steps depend heavily on it. In the initial part of pre-processing phase, the noise removal is performed by applying Wiener filter across 5 × 5 windows through all over the image. This step helps in the removal of smaller variations along with the blood smear images. Gaussian blurring is done for further smoothing of the image texture. If $W(x, y)$ is a smaller window of the input image and σ represents the standard deviation, then the resulting blurred image can be given by

$$G(x, y) = \frac{1}{\sigma\sqrt{2\pi}} e^{-(x^2+y^2)2\sigma^2} \tag{1}$$

3.2 RBC Segmentation

Malaria blood smear can be composed of three major objects—RBC, white blood cell (WBC) and platelets. As the malaria parasite resides inside RBC, the primary aim is to separate RBCs from the whole blood smear images. The general appearance of the WBC is bluish colour with a much bigger size than the RBC. Number of WBCs present in the blood smear is also much lesser than that of RBC. The same observation can be made for platelets also but the size of platelets are smaller than that of RBCs. Two different approaches for separating RBCs from the blood smear images are taken. One way is based on textural features and the other is dependent on the intensity profiles. Entropy is the measure of randomness. In case of images, low entropy of an image segment implies smoothness with lesser intensity variations and higher entropy indicates edge regions or regions with high-intensity fluctuations. If $I_{(x,y)}$ pixel of the image has the intensity level k and P denotes the probability of occurrence of that level in a pre-defined $s \times s$ window, then entropy $H(w)$ is given by

$$H(w) = -\sum_{i=1}^{s} P(k_i)\log P(k_i) \tag{2}$$

It is evident that entropy values of the blood smear will be high only in areas of background-blood object (RBC/WBC/platelet) transition and inside the blood objects (in case of affected parasites). The higher entropy regions (with entropy value H_{Entropy}) are defined as areas with entropy levels having top three maximum frequencies. The reason behind choosing top three levels is that there will be approximately three main regions,

- one for the background having the highest frequency
- second highest frequency level that represents the entropy variations inside RBCs
- third highest frequency that shows the blood particle-background border

Based on this concept, the image of the blood smear is analysed for higher entropy regions. Setting the threshold for the third highest frequency level, it gives the approximate locations of blood particles. As can be seen from Table 1 images (1, 2)(d), the entropy histograms show that the choice of $3rd$ highest entropy levels is optimum for segmenting blood particles. Along with the textural analysis, intensity profile-based analysis is also done to segment RBCs from blood smear images. Table 1 images (1, 2)(f), the plot of intensity histograms give almost clear valley indicating the region of intensities of blood particle that can be used for segmentation. The result of pre-processing is shown in Table 1 images (1, 2), where (a) indicates original input image, (b) shows the extracted green channel, (c) is the result of entropy analysis, (d) gives entropy histogram, (e) is the result of entropy-based segmentation, (f) is the intensity-level histogram and (g) shows the result of segmenting (b) using Otsu [9] threshold, while (h) gives the extracted blood particles from the blood smear images. If there exists any very large object whose size is much larger than the average RBC size, then it is marked as WBC and it is not taken into account. The centroid of the rest of the extracted components are computed and a square of a box of the size of average RBC is drawn. These computed square areas are extracted from the original blood smear image. These are the probable regions that need further probing.

3.3 Parasite Detection Using Convolutional Neural Network

Convolutional neural network (CNN) is a very efficient classification tool that uses the concept of deep learning architecture. Consisting of various layers that perform dedicated activities from convolution to data processing and downsizing, etc. CNN offers very accurate predictions. Application of CNN is wide including speech recognition, text recognition to image classification [4]. Classification of a very large set of images can be done using CNN in a very efficient way. In the proposed approach, one very simple CNN structure is built for malarial parasite detection in blood smears. A brief discussion of various blocks that have been used in the proposed approach is mentioned below.

- **Input layer**: This is the input image that is sent to CNN for classification or training. In the proposed approach, the greyscale image is used for the input layer.
- **Convolution layer**: This is the main stage of a CNN. A set of filters are present in this layer that attempts to find specific features in the image. This layer produces a dot product between the input image and the set of filters to generate activation maps that will be used in the subsequent stages. In the proposed method, convolution filters of 3×3 and 5×5 are being used.
- **Activation function**: The output generated in the convolution layer is passed to the activation layer. The decision of firing a neuron is taken by the activation layer.

Table 1 Result of pre-processing

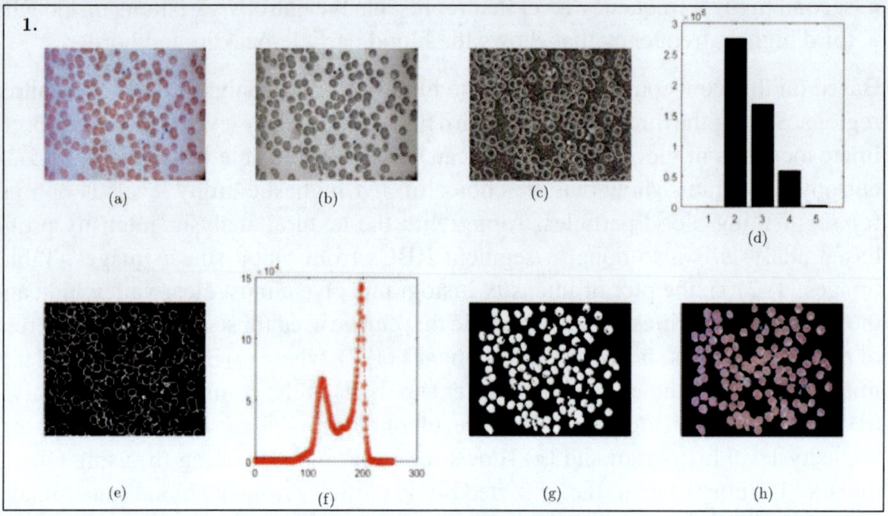

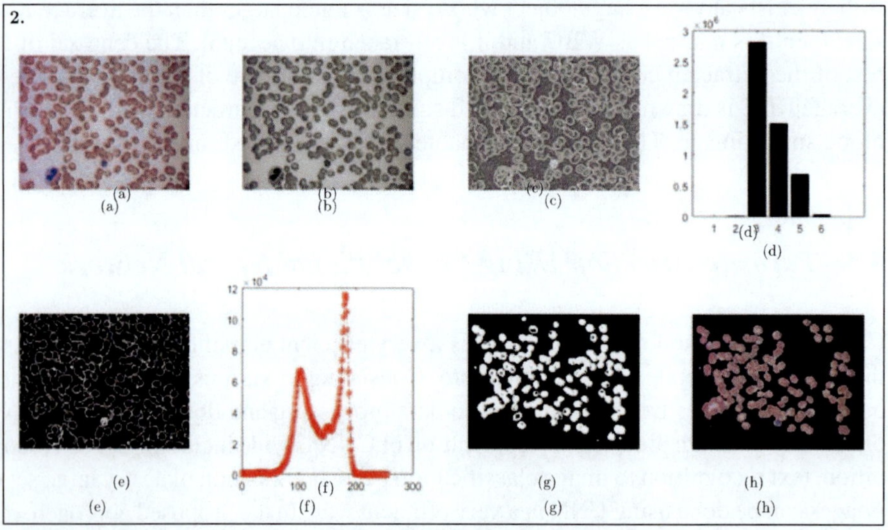

Examples are Relu, TanH, sigmoid, etc. Here, rectified linear unit (ReLU), is being used four times. The working of this function is based on the idea of selecting the maximum value (max (0, $value$))

- **Pooling layer**: This layer is responsible for the reduction of spatial image size and lesser computation cost. In the proposed method, max pooling is used. Max pooling selects the highest value in the defined neighbourhood ignoring the rest of the values. The final output of this layer is solely dependent on the size and stride of the pooling window size. Max pooling of 2×2 size with stride 2 is used in this CNN architecture.

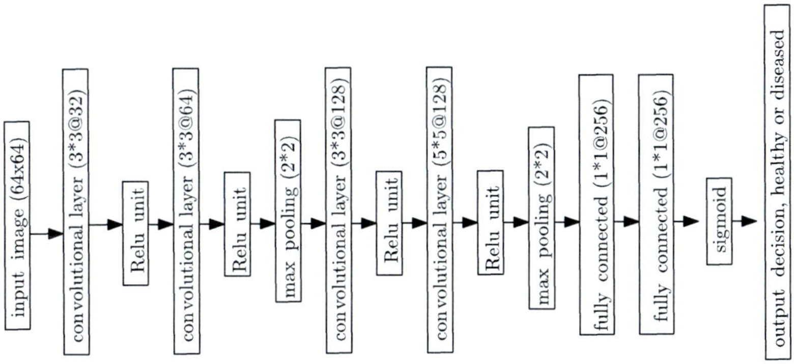

Fig. 3 Proposed CNN architecture

– **Fully connected layer**: This layer is added as the last stage of CNN. These are nothing but normal neural network layers that receives activation values from the previous layers and use it as an array of values after flattening it. Here, two fully connected layers are used to come to a final conclusion regarding the prediction of disease.

The proposed CNN architecture is depicted in Fig. 3. The performance and accuracy of the classification are completely determined by the architecture of the CNN. There are 11 layers comprising of convolution layers, Relu layer and pooling layers. To come to a decision, two fully connected layers along with sigmoid function are used.

In this work, two main datasets are used in combination for training and testing purpose. The first one is from [5] and the second one is from https://lhncbc.nlm.nih.gov/publication/pub9932. The images retrieved from the first one is segmented and the individual RBCs are extracted for training and testing purpose. Moreover, the data from second one, already a collection of individual RBCs labelled as healthy or diseased, is used in the training phase. The whole dataset is separated into 80%

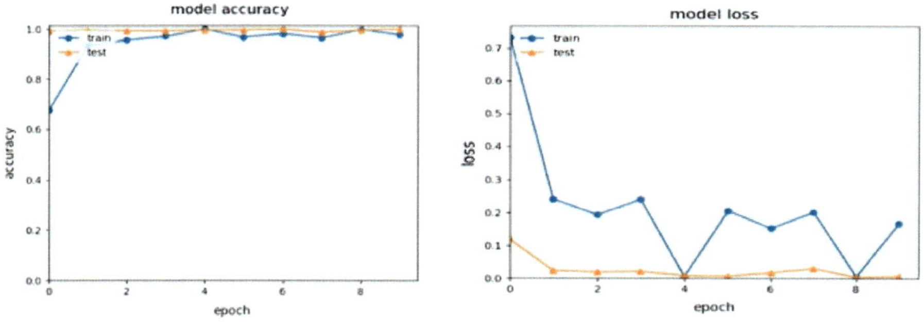

(a) Proposed model accuracy during training and testing

(b) Proposed model loss during training and testing

Fig. 4 Accuracy and loss of the CNN model

images for training purpose and the rest 20% images for testing. A ten-fold cross-validation method is applied on the dataset for evaluation of the proposed CNN model. As shown in Fig. 4a, b, the accuracy of the train and test both are high and the loss amount is also very minimum with the proposed CNN architecture. With this setup, a dice score of 0.95 is obtained using this CNN-based approach for malaria parasite detection.

4 Conclusion and Discussion

This paper attempts to classify thin blood smears based on the appearance of malaria parasites. Basic image pre-processing followed by custom CNN-based classification gives dice score of 0.95 success rate in the detection of malaria-affected blood smear images. In future, the proposed approach can be extended for classification of malarial parasites erythrocytic stages that will give deeper insight into the stage of the affected patient that will be helpful for planning treatment in a much better way. Another categorization of parasite type, whether falciparum or vivax, etc, may be made in future based on this analysis by examining the appearance of the detected parasites.

References

1. Jan, Z., Khan, A., Sajjad, M., Muhammad, K., Rho, S., Mehmood, I.: A review on automated diagnosis of malaria parasite in microscopic blood smears images. Multimed. Tools Appl. **77**(8), 9801–9826 (2018)
2. Kaur, D., Walia, G.K.: Edge detection of malaria parasites using ant colony optimization. In: 4th International Conference on Signal Processing, Computing and Control (ISPCC), pp. 451–456. IEEE (2017)
3. Khan, N.A., Pervaz, H., Latif, A.K., Musharraf, A., et al.: Unsupervised identification of malaria parasites using computer vision. In: 11th International Joint Conference on Computer Science and Software Engineering (JCSSE), pp. 263–267. IEEE (2014)
4. LeCun, Y., Haffner, P., Bottou, L., Bengio, Y.: Object recognition with gradient-based learning. In: Shape, Contour and Grouping in Computer Vision, pp. 319–345. Springer, Berlin (1999)
5. Loddo, A., Di Ruberto, C., Kocher, M., ProdHom, G.: Mp-idb: The malaria parasite image database for image processing and analysis. In: Sipaim–Miccai Biomedical Workshop, pp. 57–65. Springer, Berlin (2018)
6. Mashor, M., Nasir, A.A., Mohamed, Z., et al.: Identification of giemsa staind of malaria using k-means clustering segmentation technique. In: 6th International Conference on Cyber and IT Service Management (CITSM), pp. 1–4. IEEE (2018)
7. Moody, A.: Rapid diagnostic tests for malaria parasites. Clin. Microbiol. Rev. **15**(1), 66–78 (2002)
8. Nanoti, A., Jain, S., Gupta, C., Vyas, G.: Detection of malaria parasite species and life cycle stages using microscopic images of thin blood smear. In: International Conference on Inventive Computation Technologies (ICICT), vol. 1, pp. 1–6. IEEE (2016)
9. Otsu, N.: A threshold selection method from gray-level histograms. IEEE Trans. Syst. Man Cybern. **9**(1), 62–66 (1979)

10. Pattanaik, P.A., Swarnkar, T., Sheet, D.: Object detection technique for malaria parasite in thin blood smear images. In: IEEE International Conference on Bioinformatics and Biomedicine (BIBM), pp. 2120–2123. IEEE (2017)
11. Peñas, K.E.D., Rivera, P.T., Naval, P.C.: Malaria parasite detection and species identification on thin blood smears using a convolutional neural network. In: IEEE/ACM International Conference on Connected Health: Applications, Systems and Engineering Technologies (CHASE), pp. 1–6. IEEE (2017)
12. Poostchi, M., Silamut, K., Maude, R.J., Jaeger, S., Thoma, G.: Image analysis and machine learning for detecting malaria. Transl. Res. **194**, 36–55 (2018)
13. Reni, S.K., Kale, I., Morling, R.: Analysis of thin blood images for automated malaria diagnosis. In: E-Health and Bioengineering Conference (EHB), pp. 1–4. IEEE (2015)
14. Rosado, L., da Costa, J., Elias, D., Cardoso, J.: Mobile-based analysis of malaria-infected thin blood smears: automated species and life cycle stage determination. Sensors **17**(10), 2167 (2017)
15. Somasekar, J., Reddy, B.E.: Contrast-enhanced microscopic imaging of malaria parasites. In: IEEE International Conference on Computational Intelligence and Computing Research, pp. 1–4. IEEE (2014)
16. Wattana, M., Boonsri, T.: Improvement of complete malaria cell image segmentation. In: Twelfth International Conference on Digital Information Management (ICDIM), pp. 319–323. IEEE (2017)

A Deep Learning Approach for Predicting Air Pollution in Smart Cities

Banani Ghose and Zeenat Rehena

Abstract In recent times, air pollution has been increased in many folds due to drastic urbanization, industrial advancement, emission from the vehicles, and unrestricted destruction of the greenery in smart cities. Air pollution is having adverse side effects, which may be of long term or of short term, but in any way, air pollution is a threat to all ages of human beings. In many cases, it is out of the scope of an individual to solve the problem as a whole, but if it is possible to predict the air pollution beforehand, then many times the worst air pollution effects can be avoided. Using the predicted result, the environment management authorities can make some preventive decisions, which in turn, will help the common citizens to get rid of the problem. In this paper, a novel deep learning model has been proposed to forecast the pollution. The comparison among the existing state-of-the-art methods is also done here. Experimental results shown in this work claim that the proposed work outperforms other methods with respect to accuracy and stability.

Keywords Smart city · Deep learning · Air pollution · Prediction · LSTM

1 Introduction

Urbanization, industrialization, misutilization, etc., of fuel are the key issues which are causing air pollution in smart cities. Air pollution may be of two types, outdoor air pollution and indoor air pollution [1]. Both are having a severe impact on human lives. Air pollution affects the citizens of smart cities. The effect may be of two types: (i) Long-term effect and (ii) Short-term effect [1]. Heart diseases, lung cancer, and respiratory diseases such as emphysema are long-term effects. The citizens may

B. Ghose (✉)
Haldia Institute Technology, Haldia 721657, India
e-mail: bananiritu@gmail.com

Z. Rehena
Aliah University, Kolkata 700160, India
e-mail: zeenatrehena@yahoo.co.in

J. K. Mandal et al. (eds.), *Computational Intelligence and Machine Learning*, Advances in Intelligent Systems and Computing 1276,
https://doi.org/10.1007/978-981-15-8610-1_4

have brain, kidney, liver, or nervous system damage due to severe air pollution. The newborn may have some permanent birth defects. On the other hand, several discomforts are observed among human lives which may be of short term, which are caused by the effect of air pollution.

In the context-aware approach, many sensors are installed in the environment for capturing the air samples. These sensors are the context providers. When the air samples are collected from the air, then it is to be analyzed and the Air Quality Index (AQI) is calculated. Observing the AQI, a conclusion is to be framed out, whether the air is polluted or not. According to that analysis, the alert may be generated. So, if the citizens are informed about air pollution in advance, then they may plan to avoid the pollution prone area. Moreover, some regulatory bodies may take some measuring steps to prevent pollution. Here comes the importance of the *prediction*. If the future can be predicted, then many times using context-aware computing the alert can be generated to make the recipient aware of the fact.

There are many models available through which the forecasting may be done. Artificial neural network (ANN) has some models such as Backpropagation neural network model [2], Recurrent neural network model [3], and so on. Among all these, RNN offers so many advantages that are far better than any statistical models in terms of performance. RNN delivers forecasting results more accurately and more efficiently than the other existing models [4, 5]. For forecasting time series data, many RNN models are proposed [6, 7]. RNN also sufferers from vanishing gradient problem and thus it cannot be worked properly.

To overcome the drawbacks of RNN, Long Short-Term Memory (LSTM) was introduced. Shallow LSTM is inefficient to handle nonlinear and Long-term time series data, whereas deep LSTM with the stacking of multiple layers shows a better result compared to shallow LSTM [8, 9]. Deep LSTM with a stack of multiple layers can manage long-term data even if they are nonlinear.

In this paper, deep learning-based model by stacking of multiple LSTM layers are proposed, and that is named after as DLSTM (Deep Long Short-Term Memory) model. In this work, a performance comparison among the already existing models such as Support Vector Regression (SVR) [10], Autoregressive integrated moving average (ARIMA) [11], and shallow LSTM is also done, which shows that the DLSTM model is giving a better result with more accuracy.

2 Literature Review

To formulate this work, a number of related papers are consulted. Some of them have predicted one of the pollutants of the sample air, some of them have predicted by proposing various models. In [12], it's been seen that the transboundary problem, which is relevant for the coastal area or island areas is addressed. The authors had suggested a convolution recurrent neural network (CRNN) for considering the spatial–temporal feature of the air sample of those areas. Among so many agents in the urban air, some of the agents are mainly responsible for the unhealthy air in smart cities. $PM_{2.5}$ is one of the polluting agents. Deep Learning approach is used to build an LSTM model to overcome the air pollution problem is proposed in [13]. A comparative study is also carried out with SVR here. To forecast the $PM_{2.5}$ in air, a model is proposed using LSTM using 30 days' time series dataset in [14]. The authors compared linear regression and recurrent neural network models for the performance evaluation of LSTM. In [15], air pollution is being predicted beforehand by proposing a bidirectional LSTM model. In [16], the air pollutants are quantified using a forward deep learning approach. The future air pollution prediction model using deep learning-based LSTM is also proposed in [17]. They have also compared their prediction with the data available in the Central Pollution Control Board (CPCB). In [18], authors have also proposed a deep learning-based model to predict air pollution in South Korea using LSTM.

3 Proposed Prediction Method

The detail description of the proposed method is presented in this section. The basic building blocks of the proposed model are LSTM recurrent neural network. LSTMs have overcome the shortfall of the standard RNNs. In this work, the standard and basic LSTM architecture are deployed. The basic LSTM architecture is depicted in Fig. 1. The LSTM has cell and gates. The three gates are input gate, output gate, and forget gate, which regulates the flow of information to and from the cell.

3.1 Deep Long Short-Term Memory

It has been seen [19] that the more the number of hidden layers, the better will be the performance of the deep learning model. Therefore, in this work, a DLSTM method has been proposed. It is constructed by stacking of multiple LSTM layers as shown in the Fig. 2. Here, the output of one layer is treated as the input of the next layer.

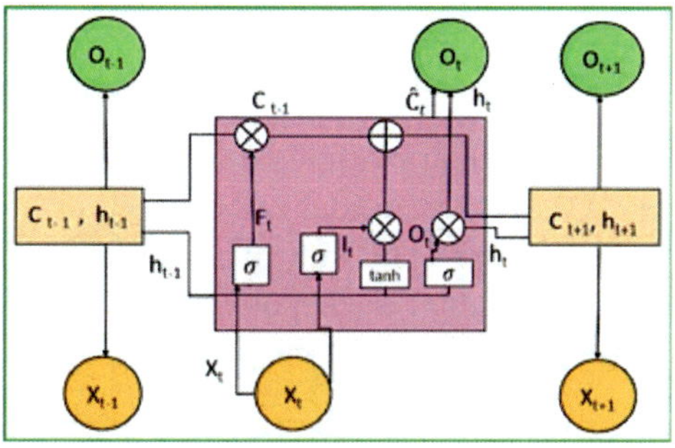

Fig. 1 Basic LSTM architecture

Fig. 2 DLSTM architecture

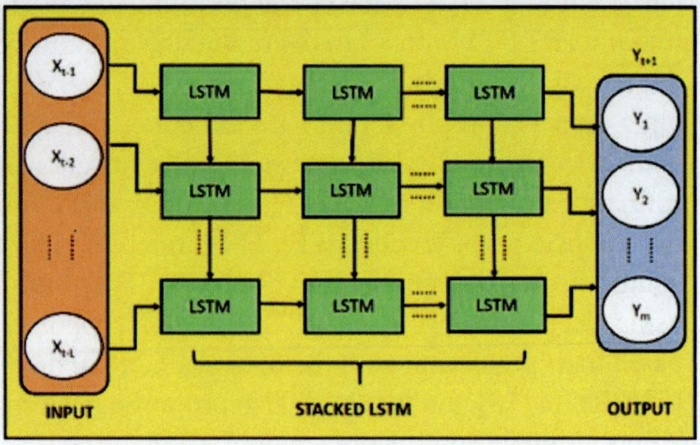

In this work, the aim is to forecast the AQI at the $(t+1)$ timestamp based on the previous n number of timestamps.

Here, x_t, x_{t-1}, ..., x_{t-n} are taken as the input of the DLSTM model and y_{t+1} is the output of the proposed model. Here, each input x_i is itself a vector of length k, $x_i = \{x_{i,1}, x_{i,2}, \ldots, x_{i,k}\}$.

3.2 Data Collection

For this work of prediction, the proposed model (DLSTM) is to be trained with some historical data. The dataset which is used in the experiment is a historical time series data [20]. The dataset comprising of the concentration of five pollutants, like $PM_{2.5}$, PM_{10}, NO_2, O_3, and SO_2 for a period of 10 years, the data is collected in each hour. The dataset has a concentration of five pollutants from 01 January 2008, 00:00 h to 31 December 2018, 23:00 h. The dataset can be termed as multivariate time series data.

3.3 AQI Calculation

AQI calculation is done based on the historical dataset [21].

- The concentration of the pollutants are the Input dataset, let it be x
- AQI of the next hour is the target, i.e., it is to be evaluated, let it be y
- So, the system is now in a position, such that, the system will be trained with the concentration of the pollutants at the time instance t and the proposed model will predict the AQI for the time interval $(t+1)$

3.4 Auto Correlation Coefficient Calculation

Auto correlation coefficient is evaluated to determine the degree of correlation within the dataset over a period of time. It is generally used to observe if there is a repetition in the pattern over the period of time. Autocorrelation coefficient is used in the perspective to determine whether there is any influence of the historical data on the current data.

- For the time slice x_t, if the correlation coefficient between x_t and x_{t-k} is determined, then that will be the autocorrelation coefficient with an interval k of x_t. The formula for autocorrelation coefficient is as follows:

$$\rho = \frac{Cov(x_t, x_{t-1})}{\sqrt{Var(x_t)Var(x_{t-k})}}; k = 1, 2, \ldots \tag{1}$$

Where, $Cov(x_t, x_{t-k})$ is autocovariance, $Var(x_t)$ and $Var(x_{t-k})$ are variances.

4 Experimental Environment

The flowchart of the experimental process is displayed in Fig. 3. Python programming language is used for all the experiments. Here, the number of LSTM layers is set to 4. The *softmax* activation function is used over here. Moreover, each layer is expanded by up to n numbers. In this model, the variable input batch sizes are used. ADAM optimizer is used in this proposed work. The window stride length is kept as 1 h. The broad steps mentioned in Fig. 3 are as follows:

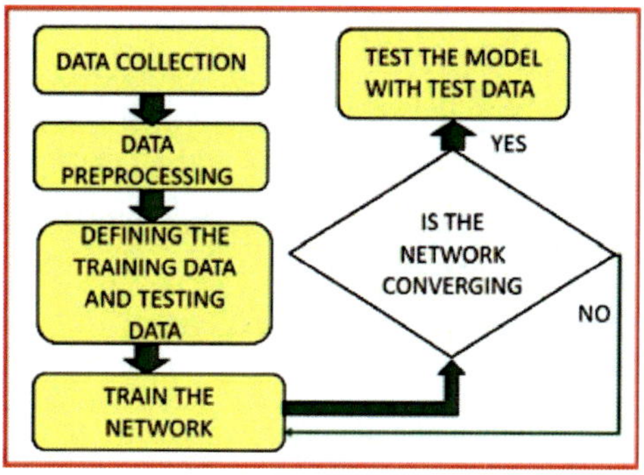

Fig. 3 Workflow structure of DLSTM model

4.1 Data Collection

The dataset is collected from London Datastore website [20]. x is a vector of length k, where k is the number of pollutants and y is the AQI to be predicted. Based on the previous n hours' data, the prediction of AQI of the $(n + 1)$th hour is done in this work. Here, the dataset has two parts, one is the training data, which is consisting of the 80% of the total dataset and, second is, the testing dataset, which is 20% of the collected dataset.

4.2 Data Preprocessing

Data preprocessing is very important for any prediction-related works. In this proposed model, the collected dataset is to be preprocessed. The first phase of this preprocessing stage involves data cleaning. In this phase, data cleaning is done to add the missing data and to discard any noisy data. Here, the missing data are replaced by the previously recorded data. The second phase of data preprocessing involves integration, as data is collected from different sources. Lastly, the normalization phase. Here, the data are so normalized that the values lie into the range of [0, 1].

4.3 Training Methodology

The proposed DLSTM model is trained with the 24 h data to make the prediction for the 25th hour. Here, ADAM (Adaptive Moment Estimation) optimizer is used to get information about network parameters. To deal with the overfitting problem, here $k - fold$ cross-validation is used to marking the stopping condition of the training process.

To train the network, the batch size is changed from time to time. The batch size varied among 4, 8, 12, 16, and 20 as shown in Table 2.

5 Results and Discussion

In this work, the air sample is collected for a long 10 years. It includes the concentration of various pollutants. The dataset consists of the concentration of $PM_{2.5}$, PM_{10}, NO_2, O_3, and SO_2 from 01 January 2008, 00:00 h to 31 December 2018, 23:00 h. Here, the DLSTM model is trained with the dataset for 24 h and the prediction for the 25th hour is aimed. In this work, the autocorrelation of AQI in every hour is calculated. It is observed in Fig. 4 that the pattern of the graph is following the same

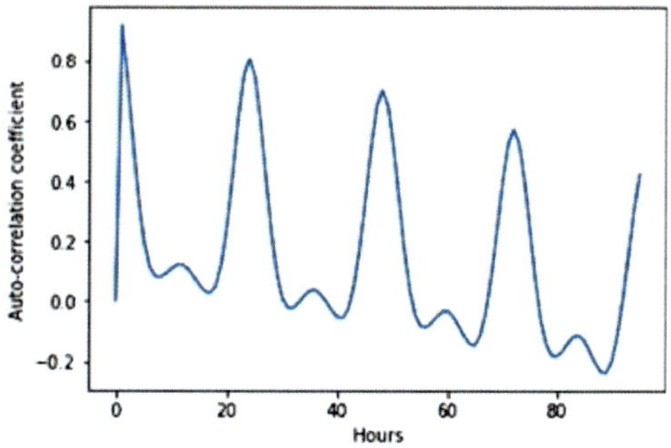

Fig. 4 Autocorrelation coefficient of hourly AQI

Table 1 Formula used for estimating errors

Error	Formula	Error	Formula
Error at time t	$e_t = y_t - \hat{y}_t$	MSE	$\frac{1}{n}\sum_{t=1}^{n} e_t^2$
RMSE	$\sqrt{\frac{1}{n}\sum_{t=1}^{n} e_t^2}$	MAE	$\frac{1}{n}\sum_{t=1}^{n} \lvert e_t \rvert$
Where y_t is actual	Value and $\hat{y}_t$ is predicted	value at the timestamp t	

Table 2 DLSTM performance evaluation in different batch sizes and epochs

Batch size	Epoch	MSE	RMSE	MAE	MSE	RMSE	MAE
4	110	0.0064	0.0800	0.0595	0.0051	0.0717	0.0532
8	201	0.0062	0.0787	0.0576	0.0050	0.0711	0.0526
12	297	0.0058	0.0761	0.0514	0.0048	0.0693	0.0491
16	**142**	**0.0047**	**0.0685**	**0.0416**	**0.0020**	**0.0451**	**0.2416**
20	197	0.0049	0.0700	0.0421	0.0021	0.0461	0.0272

pattern in every 24 h. That is the reason why the concentration of the air pollutants for the 24 h is taken as the input of the model to predict the next hours' AQI.

In this paper, the proposed model is compared with the existing baseline models, such as SVR, ARIMA, Stacked-RNN, and LSTM. Here, to compare the effectiveness of the proposed model, the Mean Square Error(MSE), Root Mean Square Error(RMSE), and Mean Absolute Error(MAE) are used. Formula to calculate these errors is presented in Table 1.

The performance of DLSTM model is presented in Table 2 with the different training batch sizes. It is clearly observed from the table, the DLSTM model produces low training and testing errors for all the batch sizes. When the training batch size is 16 the proposed model exhibits its best performance compared to other training bach sizes.

Observing the output of the proposed model in Table 3, it can be easily concluded that, the model is outperforming the other baseline prediction models.

Table 3 Performance evaluation of existing model and proposed DLSTM model

Model	MSE	RMSE	MAE
SVR	0.0053877	0.0734000	0.0569350
ARIMA	0.0020598	0.0453850	0.0301014
DLSTM	**0.0020416**	**0.0451840**	**0.0241671**
RNN	0.0031743	0.0563400	0.0336400
LSTM	0.0028975	0.0538280	0.0308200

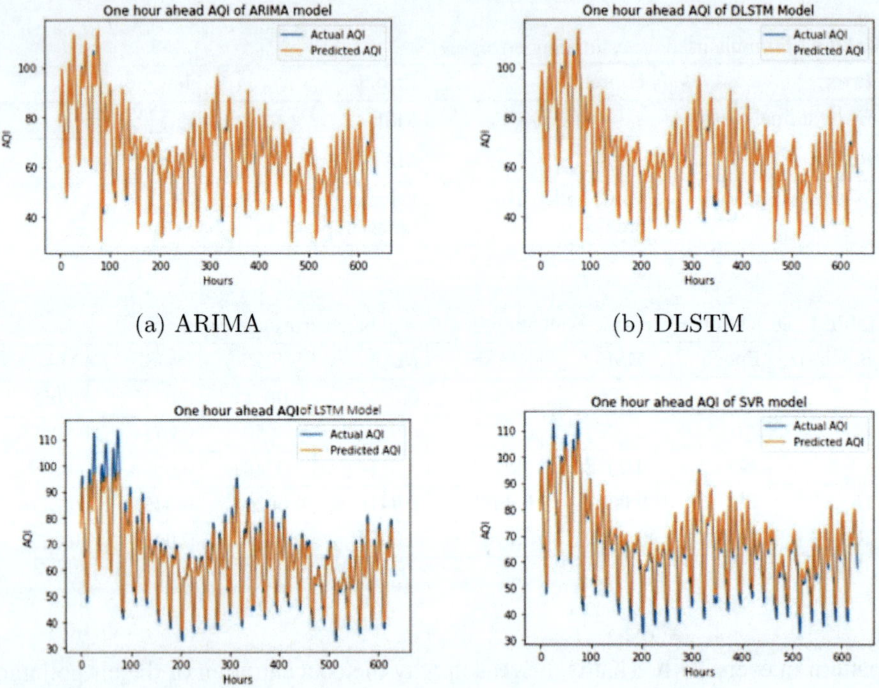

(a) ARIMA (b) DLSTM

(c) LSTM (d) SVR

Fig. 5 Comparison of actual and predicted AQI of different models. **a** ARIMA Model, **b** DLSTM Model **c** Shallow LSTM Model and **d** SVR Model

In Fig. 5, the prediction result of every baseline models, like ARIMA, shallow LSTM and SVR are shown in (a), (c), and (d), respectively, whereas in (b), the graphical representation of the prediction of the proposed model DLSTM is shown. From these graphs, it is clearly visible that, the proposed DLSTM model outperforms the older models.

Observing the experimental result in Fig. 4, it can be vividly said that the hourly AQI is repeating the similar pattern in a fixed time interval. That means, knowing the past AQI pattern, what should be the AQI of the next hours that can be easily predicted.

6 Conclusion

Air pollution is a challenge which can be addressed in many ways, but, if the possibility of air pollution can be sensed beforehand, the ill effects of air pollution can be avoided in many cases. There lies that importance of prediction. The context-aware computing approach to handle air pollution in smart cities is a work which will be helpful in the day to day lives of the citizens. In the DLSTM model, it is shown that the prediction of the possibility of air pollution can be efficiently forecasted. The autocorrelation coefficient will also help in predicting the AQI by having knowledge of the previous hours' AQI. Using these results, in future, the alert generation will be easier in context aware computing approach for handling air pollution. In future, the prediction may be refined by making the model more sensible and efficient.

References

1. WHO: Air Pollution (2019). https://www.who.int/airpollution/en/
2. Law, R.: Back-propagation learning in improving the accuracy of neural network-based tourism demand forecasting. Tourism Manag. **21**(4), 331–340 (2000)
3. Hüsken, M., Stagge, P.: Recurrent neural networks for time series classification. Neurocomputing **50**, 223–235 (2003)
4. Zhang, G., Patuwo, B.E., Hu, M.Y.: Forecasting with artificial neural networks: the state of the art. Int. J. Forecat. **14**(1), 35–62 (1998)
5. Connor, J.T., Martin, R.D., Atlas, L.E.: Recurrent neural networks and robust time series prediction. IEEE Trans. Neural Netw. **5**(2), 240–254 (1994)
6. Barbounis, T.G., Theocharis, J.B., Alexiadis, M.C., Dokopoulos, P.S.: Long-term wind speed and power forecasting using local recurrent neural network models. IEEE Trans. Energy Convers. **21**(1), 273–284 (2006)
7. Schuster, M., Paliwal, K.K.: Bidirectional recurrent neural networks. IEEE Trans. Signal Process. **45**(11), 2673–2681 (1997)
8. Sagheer, A., Kotb, M.: Time series forecasting of petroleum production using deep lstm recurrent networks. Neurocomputing **323**, 203–213 (2019)
9. Zhao, Z., Chen, W., Wu, X., Chen, P.C., Liu, J.: Lstm network: a deep learning approach for short-term traffic forecast. IET Intell. Transp. Syst. **11**(2), 68–75 (2017)
10. Smola, A.J., Schölkopf, B.: A tutorial on support vector regression. Stat. Comput. **14**(3), 199–222 (2004)
11. Box, G.E., Pierce, D.A.: Distribution of residual autocorrelations in autoregressive-integrated moving average time series models. J. Am. Stat. Assoc. **65**(332), 1509–1526 (1970)
12. Zhao, P., Zettsu, K.: Convolution recurrent neural networks based dynamic transboundary air pollution predictiona. In: 2019 IEEE 4th International Conference on Big Data Analytics (ICBDA), pp. 410–413. IEEE (2019)
13. Kök, İ., Şimşek, M.U., Özdemir, S.: A deep learning model for air quality prediction in smart cities. In: 2017 IEEE International Conference on Big Data (Big Data), pp. 1983–1990. IEEE (2017)
14. Park, J.H., Yoo, S.J., Kim, K.J., Gu, Y.H., Lee, K.H., Son, U.H.: Pm10 density forecast model using long short term memory. In: 2017 Ninth International Conference on Ubiquitous and Future Networks (ICUFN), pp. 576–581. IEEE (2017)
15. Verma, I., Ahuja, R., Meisheri, H., Dey, L.: Air pollutant severity prediction using bi-directional lstm network. In: 2018 IEEE/WIC/ACM International Conference on Web Intelligence (WI), pp. 651–654. IEEE (2018)

16. Rao, K.S., Devi, G.L., Ramesh, N.: Air quality prediction in visakhapatnam with lstm based recurrent neural networks. Int. J. Intell. Syst. Appl. **11**(2), 18 (2019)
17. Chaudhary, V., Deshbhratar, A., Kumar, V., Paul, D.: Time series based lstm model to predict air pollutants concentration for prominent cities in india (2018)
18. Bui, T.C., Le, V.D., Cha, S.K.: A deep learning approach for forecasting air pollution in south korea using lstm (2018). arXiv:1804.07891
19. LeCun, Y., Bengio, Y., Hinton, G.: Deep Learn. Nat. **521**(7553), 436 (2015)
20. London, K.C.: Air Polutant Dataset (2019). https://data.london.gov.uk/download/london-average-air-quality-levels/576175f3-b515-4c77-a6a2-6e222a3e5340/air-quality-london-time-of-day.csv
21. Board, C.P.C.: Air Quality Index (2014). http://www.indiaenvironmentportal.org.in/files/file/Air%20Quality%20Index.pdf

Structural Design of Convolutional Neural Network-Based Steganalysis

Pratap Chandra Mandal

Abstract A tailored Convolutional Neural Network (CNN) model aimed at steganalysis has been designed. The model is able to acquire the complicated dependencies which are necessary for steganalysis. The usefulness of the model has been demonstrated on the state-of-the-art method S-UNIWARD at 0.4 bpp. The model achieves F1 score equal to 0.849 and detection error 0.152 which are comparable performance.

Keywords Steganalysis · Features · CNN · S-UNIWARD

1 Introduction

Steganalysis [2, 10] has been widely used since last 20 years. It is used to identify the existence of unseen data within a cover medium. Steganalysis can be performed in two stages: feature extraction followed by classification. In Deep learning (DL), feature representations can be learned automatically. CNNs [3, 4] have got remarkable success in computer vision in last few years.

Krizhevsky et al. [3] have designed a CNN on the subsets of ImageNet dataset and achieved the best result during that time. The network has five convolutional layers (CONV) with three fully connected (FC) layers. By the introduction of several unusual new features, it reduces the training time and improves its performance. Rectified Linear Unit (ReLU) as an activation function trains the model several times faster than tanh units. Tan and Li [5] have presented a nine-layer CNN-centred blind steganalyzer. The model showed better performance than SPAM. It was the beginning of the adventure in DL based steganalysis. A few months later,

P. C. Mandal (✉)
Department of Computer Science and Engineering, B.P. Poddar Institute of Management and Technology, Kolkata 700052, India
e-mail: pcmandal9@gmail.com

Department of Computer Science and Engineering, Indian Institute of Information Technology Kalyani, Kalyani, India

J. K. Mandal et al. (eds.), *Computational Intelligence and Machine Learning*, Advances in Intelligent Systems and Computing 1276,
https://doi.org/10.1007/978-981-15-8610-1_5

Qian et al. [4] have proposed a new model, called Gaussian Neuron Convolutional Neural Network model which learns feature representations using several CONV layer. Results showed the usefulness of their model on three algorithms: S-UNIWARD, WOW and HUGO. It has achieved comparable performance with SRM. Layer 5 (last layer) provides only 256 features. But in SRM, dimension of the feature vector provides 34671. Xu et al. [8] designed a CNN containing five CONV layers with an FC layer. Experimental results confirmed that the performance of their design is as good as the performance of SRM. DL steganalyzers not only became deeper but also complex afterwards. Paper [7] showed the extended work of Xu et al. [8]. Results are comparable to SRM. Paper [9] presented a model with a truncated linear unit for boosting detection process further. The efficiency of their CNN model was further enhanced by introducing the selection channel. Paper [10] have used quantization and truncation into DL for the first time. It has proved their superiority over traditional JPEG steganalytic features. Wu et al. [6] have recommended a relatively bigger deep CNN (DRN) than existing CNNs. Its accuracy is quite higher than the other models. Paper [11] have proposed a CNN model for detection of steganography in larger JPEG images. Experimental results showed the better revealing correctness and faster convergence rate for large images.

1.1 Motivation

Steganalysis can be performed in two stages: feature extraction followed by classification. In traditional steganalysis, these two stages are not joined, simultaneous optimization is not possible. So, the supervision of classification can not be used to acquire convenient data that we get from the feature extraction stage.

To overcome the problem just mentioned, feature representations can be learned as a replacement of planning for hand-crafted features. In deep learning, feature representations can be learned automatically. CNNs have got remarkable successes in steganalysis in recent years. For this reason, we are interested in steganalysis using CNNs.

1.2 The Structure of Prevailing Steganalysis Methods

Traditional steganalysis is presented at the top of Fig. 1. The design of our model is shown below. The model comprises three portions: an image processing, multiple CONV layers with multiple FC layers. Here, the parameters are learned automatically in comparison to the traditional methods.

We have prepared the remaining part of the paper in this way: proposed work is described in Sect. 2. Section 3 explains results and discussions. Section 4 describes the conclusion and future research opportunity.

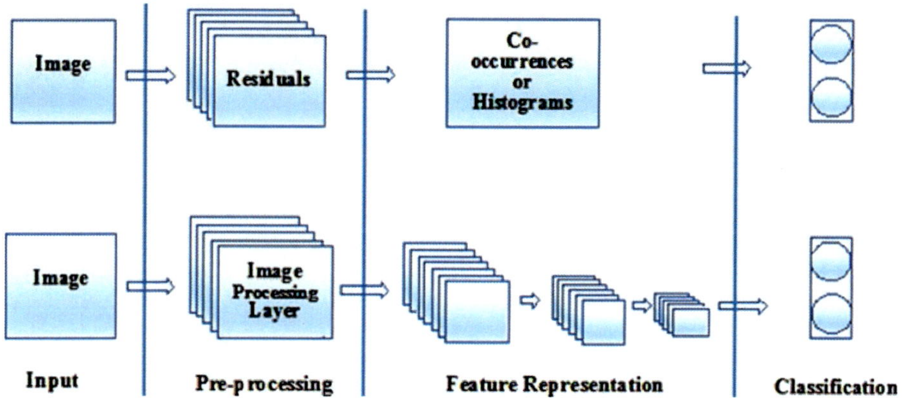

Fig. 1 Design of traditional steganalysis methods [9] with its similar CNNs

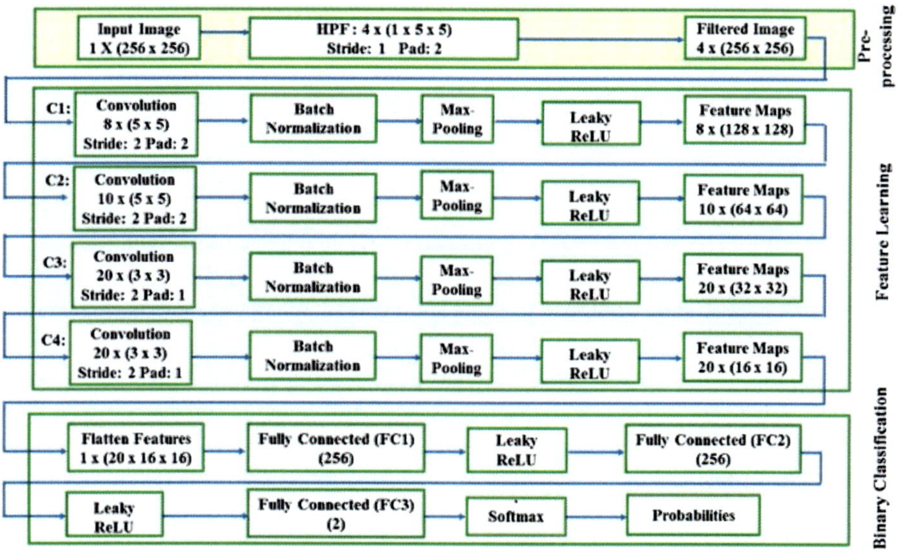

Fig. 2 Architecture of the proposed Model

2 Proposed Work

Here, a CNN-based steganalysis model in the spatial domain has been presented. At first, the model is trained using cover and stego images, then unknown images are checked for deciding the presence of the hidden information.

The model uses the input images of size 256 × 256. It comprises a pre-processing block, feature learning block and an FC block as shown in Fig. 2. Feature learning block consists of four CONV, whereas FC block consists of three dense layers. The model generates a probability distribution on two classes: cover and stego. The pre-processing block [1] is used for filtering the input images by four high-pass filters

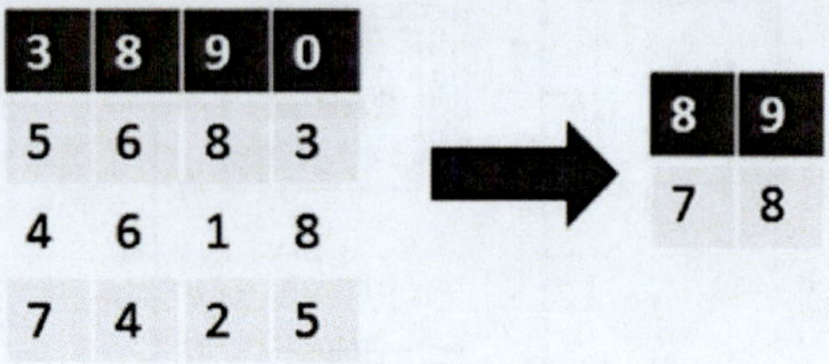

Fig. 3 Max pooling operation

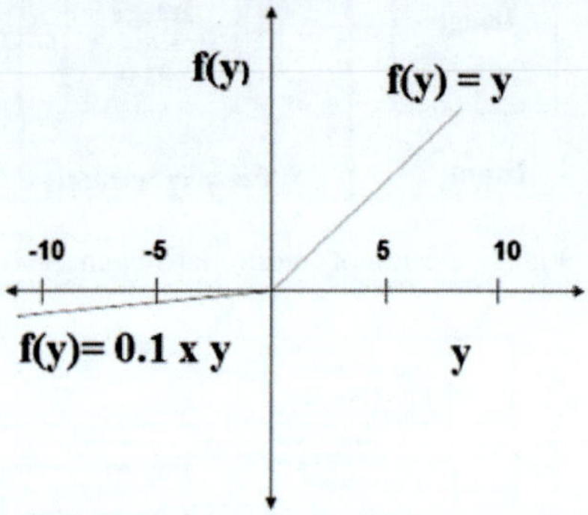

Fig. 4 Leaky ReLU activation function

for extracting the noise component residuals. The first CONV layer is formed with eight neurons. The second CONV layer is formed by ten neurons. Each neuron uses 5 × 5 kernel . Here, padding and stride set to 2. The kernel size has been set to 3 × 3 with stride 2 and padding 1 in third and fourth CONV layers that consist of 20 neurons.

A Batch Normalization (BN) layer is used for improving convergence of the network. Max pooling has been provided after a BN layer as shown in Fig. 3.

The LeaKy ReLU activation function (Fig. 4) is used for all activation layers because of its fast gradient computation. The function is as follows:

$$f(y) = \begin{cases} y, & y \geq 0 \\ 0.1 \times y, & y < 0 \end{cases} \tag{1}$$

At last, pulled out features are fed into the binary classification block which comprises three FC layers. First, two layers composed of 256 neurons to prepare flattened data to the softmax function which is used to predict the probabilities of stego and cover images. The last layer has two neurons that correspond to the networks output classes. The threshold set over here is 0.6.

3 Results and Discussions

We have used cropped BOSSBase database comprising of 10,000 pairs (cover and stego) of greyscale images of size 256 × 256. The stego images are obtained with S-UNIWARD at 0.4 bpp. Nvidia GeForce 1080TI GPU has been used for the experimental set-up. From 20,000 images, 4,000 images were used for testing purpose. Leftover 8,000 pairs were used for training and validation. We have selected 6,000 pairs as a training set.

To fully evaluate the effectiveness of a model, we must examine both recall and precision, where

$$\text{Recall} = \frac{\#\text{True Positive}}{\#\text{True Positive} + \#\text{False Negative}}$$

$$\text{Precision} = \frac{\#\text{True Positive}}{\#\text{True Positive} + \#\text{False Positive}}$$

Confusion matrix of the model is shown in Table 1.
Here,

$$\text{Recall} = \frac{1708}{1708 + 0292} = 0.854$$

$$\text{Precision} = \frac{1708}{1708 + 0316} = 0.8438$$

$$\text{F1 score of the model} = 2 \times \frac{\text{Recall} \times \text{Precision}}{\text{Recall} + \text{Pecision}} = 0.8488$$

The model shows a very competitive result.

Performance of the model can be evaluated by average detection error rate D_e as follows:

$$D_e = \frac{1}{2}(D_{md} + D_{fa}) \tag{2}$$

where D_{md} states the missed detection rate and D_{fa} states the false alert rate.

Table 1 Confusion matrix of the proposed model

	Predicted: No	Predicted: Yes	Total
Actual: No	1684	0316	2000
Actual: Yes	0292	1708	2000
Total	1976	2024	4000

Table 2 Comparison of average detection error rate with state-of-the-art models at 0.4 bpp

Steganalytic methods	S-UNIWARD (%)
Qian's network [4]	30.9
Xu's network [8]	19.7
Zhang's network [12]	15.3
Wu's network [6]	06.3
Proposed model	15.2

Table 2 demonstrates the performance comparison of our model with four CNN models. It shows that our model is superior than three other models. Only the fourth model shows better result than ours, but it requires much higher computaional time due to its deep architecture.

4 Conclusion and Future Research Opportunity

We have designed a tailored CNN model aimed at steganalysis. Strength of the model comes by using four high-pass filters for residual noise extraction. Precision and recall values are 0.843 and 0.854. F1 score is equal to 0.849 which is comparable to other CNN models. Detection error of our model is better than three other CNN models.

The model can be extended to identify the hidden information in the stego image which at present is a research problem and a lot of work has been done to extract the information.

References

1. Chen, M., Sedighi, V., Boroumand, M., Fridrich, J.: Jpeg-phase-aware convolutional neural network for steganalysis of jpeg images. In: Proceedings of the 5th ACM Workshop on Information Hiding and Multimedia Security, pp. 75–84. ACM (2017)
2. Fridrich, J., Kodovsky, J.: Rich models for steganalysis of digital images. IEEE Trans. Inf. Forensics Secur. 7(3), 868–882 (2012)
3. Krizhevsky, A., Sutskever, I., Hinton, G.E.: Imagenet classification with deep convolutional neural networks. In: Advances in Neural Information Processing Systems, pp. 1097–1105 (2012)
4. Qian, Y., Dong, J., Wang, W., Tan, T.: Deep learning for steganalysis via convolutional neural networks. In: Media Watermarking, Security, and Forensics 2015, vol. 9409, p. 94090J. International Society for Optics and Photonics (2015)
5. Tan, S., Li, B.: Stacked convolutional auto-encoders for steganalysis of digital images. In: Signal and Information Processing Association Annual Summit and Conference (APSIPA), 2014 Asia-Pacific, pp. 1–4. IEEE (2014)

6. Wu, S., Zhong, S., Liu, Y.: Deep residual learning for image steganalysis. Multimed. Tools Appl. **77**(9), 10437–10453 (2018)
7. Xu, G., Wu, H.Z., Shi, Y.Q.: Ensemble of cnns for steganalysis: An empirical study. In: Proceedings of the 4th ACM Workshop on Information Hiding and Multimedia Security, pp. 103–107. ACM (2016)
8. Xu, G., Wu, H.Z., Shi, Y.Q.: Structural design of convolutional neural networks for steganalysis. IEEE Signal Process. Lett. **23**(5), 708–712 (2016)
9. Ye, J., Ni, J., Yi, Y.: Deep learning hierarchical representations for image steganalysis. IEEE Trans. Inf. Forensics Secur. **12**(11), 2545–2557 (2017)
10. Zeng, J., Tan, S., Li, B., Huang, J.: Large-scale jpeg image steganalysis using hybrid deep-learning framework. IEEE Trans. Inf. Forensics Secur. **13**(5), 1200–1214 (2018)
11. Zhang, Q., Zhao, X., Liu, C.: Convolutional neural network for larger jpeg images steganalysis. In: International Workshop on Digital Watermarking, pp. 14–28. Springer, Berlin (2018)
12. Zhang, R., Zhu, F., Liu, J., Liu, G.: Efficient feature learning and multi-size image steganalysis based on cnn (2018). arXiv:1807.11428

Natural Language Processing

Sarcasm Detection of Media Text Using Deep Neural Networks

Omkar Ajnadkar

Abstract Sarcasm detection in media text is a binary classification task where text can be either written straightly or sarcastically (with irony) where the intended meaning is the opposite of what is seemingly expressed. Performing sarcasm detection can be very useful in improving the performance of sentimental analysis where existing models fail to identify sarcasm at all. We examine the use of deep neural networks in this paper to detect the sarcasm in social media text(specifically Twitter data) as well as news headlines and compare the results. Results show that deep neural networks with the inclusion of word embeddings, bidirectional LSTM's and convolutional networks achieve better accuracy of around 88 percent for sarcasm detection.

Keywords Sarcasm · Deep learning · Word embedding · Text classification · Attention

1 Introduction

The Free Dictionary [1] defines sarcasm as the cutting, often ironic remark intended to express contempt or ridicule. If we thought in the context of sentimental analysis, a sarcastic sentence which has positive visual sentiment is the negative sentiment in actual. Many times even humans find it difficult to judge the sentence as a sarcastic at first glance. This problem leads to automatic sarcasm detection as an essential research topic to predict sentiment in the text with perfection.

If we see the posts by people social media text, we can find many instances of sarcastic text. For example, the sentence "*I guess lots of people are being hacked. Thank you, Facebook for being so secure.*," Would be detected as positive sentiment by many automatic sentiment analyzer models due to the phrase like '*thank you*' in the sentence. On the other hand, this sentence is expressed sarcastically with the

O. Ajnadkar (✉)
Indian Institute of Information Technology, Kalyani 741235, West Bengal, India
e-mail: omkarajnadkar@gmail.com
URL: http://www.iiitkalyani.ac.in

J. K. Mandal et al. (eds.), *Computational Intelligence and Machine Learning*, Advances in Intelligent Systems and Computing 1276,
https://doi.org/10.1007/978-981-15-8610-1_6

negative sentiment implied by using irony. We try to detect such examples using our deep learning-based models.

For our experiments, we are using Twitter data as one of the datasets as it is easy to collect using Twitter APIs and is widely available for text analysis. Another reason for using Twitter data is that it is easier to detect sarcasm in short sentences (due to character limitations of Twitter) than on the news items or blog posts. We are using news headlines dataset to compare how the language structure affects the sarcasm detection accuracy. We used this already available datasets for the sarcasm detection for comparing our models score with the existing models.

Existing works on sarcasm detection are aimed mainly at using unigrams and n-grams along with semi-supervised and unsupervised methods which are based on sentimental features. Instead, we are proposing the use of deep neural networks (DNN) which learns the features automatically from the data and use them to classify text as sarcastic and non-sarcastic.

Therefore the main contributions of this paper are as follows:

- We propose an efficient way to detect sarcasm on Twitter data as well as news headlines using deep neural networks (i.e., whether the text is sarcastic or not) to enhance the accuracy of sentiment analysis.
- We compare different neural network models to see the advantages and disadvantages of various architectures.
- Finally, we compare our models with existing models to check the performance of the proposed deep learning model against existing models.

The remainder of this paper is structured as follows: Sect. 2 describes some previous related work regarding sarcasm detection. Section 3 details our proposed approach for the detection of sarcasm. In Sect. 4, we discuss the obtained experimental results using our approach and compare them with existing results, and Sect. 5 concludes this work.

2 Related Work

Automated sarcasm detection is the next step in the analysis of a large amount of social media data, with sentiment analysis mainly discovered as the primary aim in the last few years. Sarcasm detection is an important step in sentiment analysis, considering its difficulty to analyze in the text. Previous works can be classified mainly in two categories: Rule and Feature-based approaches and deep learning-based approaches.

2.1 Rule and Feature-Based Approaches

Rule-based approaches predict sarcasm based on specific rules or patterns in the text. Maynard and Greenwood [2] propose that leading indicator of sarcasm in the

Twitter text is hashtags. Authors, most of the times, use hashtags to indicate sarcasm and difference in the sentiment of the hashtag, and the rest of the tweet indicates sarcasm. Riloff et al. [3] state sarcasm to be a contrast to positive sentiment words and negative situations. Ptáček et al. [4] worked on Twitter datasets in Czech and English languages for sarcasm detection. They used maximum entropy followed by SVM classifier to achieve F1-score of 0.946 on balanced and 0.924 on the imbalanced dataset. Joshi et al. [5] use multiple features comprising lexical, pragmatics, implicit, and explicit context incongruity. In the specific case, they include relevant features to detect the sentimental expectations in the sentence. For implicit incongruity, they generalize Riloff et al. [3] by identifying verb–noun phrases containing contrast in both polarities. In the feature-based approach, the focus is on the text-based features such as punctuations and sarcastic patterns in Tsur et al. [6]; N-grams, emotion marks, and intensifiers in Liebrecht et al. [7]; Unigrams, Implicit and explicit incongruity-based features in Joshi et al. [5].

2.2 Deep Learning-Based Approaches

Deep learning-based approaches have become quite popular in recent years due to a large amount of data available for sarcasm detection. The similarity between word embeddings is used for sarcasm detection in Joshi et al. [8]. The combination of the convolutional neural network and LSTM followed by a DNN is used by Ghosh and Veale [9]. The sentiment, emotion, and personality-based features are extracted to detect sarcasm using a pre-trained convolutional neural network in Poria et al. [10]. They used features extracted by CNN to perform classification by SVM on three different datasets. We also followed a deep learning-based approach due to the following reasons:

- Deep Neural networks can learn and model non-linear and complex relationships.
- Deep Neural networks can generalize better than traditional machine learning models after learning from data and predicting for unseen data.
- Combination of various deep learning-based approaches like CNNs, RNNs, and LSTMs have been successful in many NLP tasks in recent years. as well.

3 The Proposed Framework

3.1 Data

For training and testing our approach, we used the following two datasets obtained from separate sources.

Table 1 Twitter dataset used for the paper

Train set		Test set	
Sarcastic	Non-sarcastic	Sarcastic	Non-sarcastic
24,453	26,736	1,419	2,323
51,189		3,742	

Table 2 News headlines dataset used for the paper

Train set		**Test set**	
Sarcastic	Non-sarcastic	Sarcastic	Non-sarcastic
9,379	11,988	2,345	2,997
21,367		5,342	

Social Media Text (Twitter) This dataset is collected by Aniruddha Ghosh and Tony Veale for the paper [11]. This dataset is split into two parts: the train set and the test set. Train set which contains 51,189 tweets out of which 24,453 are sarcastic and remaining 26,736 are non-sarcastic. The test set contains 3,742 tweets out of which 1,149 are sarcastic and remaining 2,323 are non-sarcastic (Table 1).

News Headlines As Twitter datasets face limitations in natural language processing tasks as noise is present due to informal style of writing, this News Headlines dataset [12] for Sarcasm Detection is collected from two news website. The Onion aims at producing sarcastic versions of current events, so all the headlines from News in Brief and News in Photos categories (which are sarcastic) are collected from it. Real (and non-sarcastic) news headlines are collected from HuffPost. As the dataset is not split explicitly into train and test set, we distribute the data into an 80:20 ratio for the experimentation purpose. Train set which contains 21,367 headlines out of which 11,988 are non-sarcastic and remaining 9,379 are sarcastic. The test set contains 5,342 headlines out of which 2,997 are non-sarcastic and remaining 2,345 are sarcastic (Table 2).

We used only training set data to train our model and test it on test sets of the dataset separately to check the performance of our proposed model.

3.2 Data Preprocessing

Data obtained from Twitter is noisy due to many reasons such as text shortcuts used by people, spelling mistakes done while quick typing and grammatical errors. This informal style of writing is quite different from news headlines dataset, where grammatical and spelling checking is done before publishing the news and thus making it less prone to errors. This is why preprocessing is an essential step before applying any further model for sarcasm prediction. We perform several steps in series on the Twitter dataset to make data clean for our model as follows:

- HTML Decoding
- UTF-8 Byte Order Marking
- Removing '@' Mentions
- Removal of URL Links
- Removal of Hashtag symbols keeping words
- Expanding the Contracted Words
- Spelling Correction

3.3 Feature Extraction

After the pre-processing, it is essential to analyze the data using the hand-crafted features obtained from the text. We mainly focused on the features to get a basic understanding of the text and how various parameters change for sarcastic and non-sarcastic text. We use these features along with features extracted from the attention layer as input to the convolutional layer.

Text Based Features Text-based features are the features which are directly based on the grammatical structure of the sentence. These features include punctuation symbols and various part of speech tags of words. Specifically, we derived the following features from the text:

- Count of nouns and verbs
- Count of punctuation symbols—Question marks and exclamation marks
- Count of uppercase words
- Count of interjections

Emotion-Based Features The emotion or the sentiment of the words give us more context about the text. Hence, we extracted some features based on the sentiment and intensifiers present in the text:

- Sentiment score of text
- Positive and negative words count
- Positive and negative intensifiers count
- Count of total polarity flips in a sentence
- Bigrams and trigrams in a sentence related to sarcasm class

3.4 Proposed Deep Neural Network

The proposed deep neural network contains nine different layers: the input layer, word embedding layer, bidirectional LSTM layer, bidirectional CuDNNLSTM layer, the

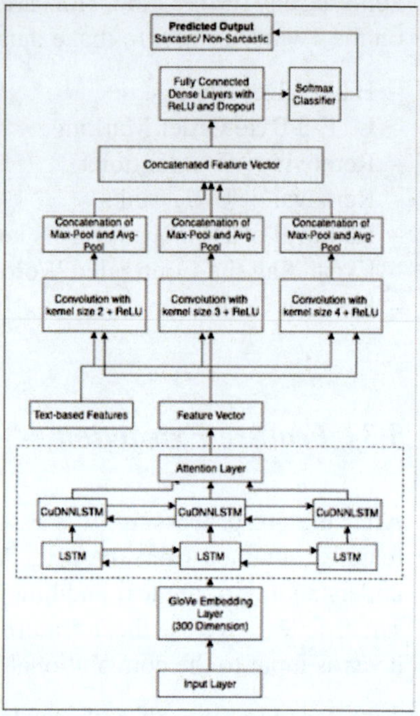

Fig. 1 Proposed Model Architecture

Attention layer, 1D convolutional layer, the dropout layer, max, and average pooling layers, fully connected dense layer, activation layer, and final representation layer. Figure 1 shows the architecture of the proposed model. The proposed model uses text and emotion-based features along with features extracted from layers of the deep neural network based on the attention mechanism to make it more generalized and robust.

Input Layer After performing all the pre-processing steps mentioned in earlier sections, the text is fed into the input layer. This layer is the starting point of the model and is connected to the next embedding layer, which builds embeddings for words in the text using the pre-trained word embeddings.

Embedding Layer Previous approaches to natural language processing problems were to create one-hot encoded vectors which are high-dimensional and sparse. Word embeddings are used to avoid this computationally inefficient approach. In this paper, to build word embeddings, we use pre-trained GloVe (Global Vectors for Word Representation) [13] embeddings of 300 dimensions which are trained on Wikipedia 2014 and Gigaword 5 dataset. To create word embeddings for our dataset, corresponding to the small size, we first limited our vocabulary to 10,000 words. As the texts are of different length, the text with the maximum length in the given dataset used as the threshold. This creates zero paddings to all the texts which are smaller than the threshold value, and all the texts finally have a length equal to the threshold value. When attached with the previous input layer, this layer creates a 2D matrix of size (threshold, 300) for each text input from the pre-trained embeddings. This matrix is finally given as input to the bidirectional LSTM layer.

Bidirectional LSTM Layer LSTM (Long Short-Term Memory) is a special type of Recurrent Neural Networks (RNN) which are capable of learning long-term dependencies. All RNNs have the form of a chain of repeating modules of neural network. This repeating module in LSTM contains three gates, which are a way to let information through the LSTM module optionally. The input gate i_t, forget gate f_t, and output gate o_t. The LSTM cell takes the cell input state x_t and the previous cell output state c_{t-1} as input and output the cell output state c_t. The bidirectional LSTMs involves duplicating the first LSTM layer in the network, which creates another layer side by side. It then provides the reverse copy in the input sequence to the second LSTM layer, which provides additional context to the network [14]. This both outputs then gets concatenated to pass to the next layer.

Bidirectional CuDNNLSTM Layer CuDNNLSTM is the faster implementation of LSTM which works the same way as the above, but using NVIDIA's Deep Neural Network library (cuDNN) [15] which is a GPU-accelerated library. The output of this layer is passed as input to the next Attention layer.

Attention Layer Attention is used to alter the fixed-length representation in the encoder–decoder architecture. The idea of attention [16] is that it trains the model to learns to pay selective attention to the inputs and relate them to items in the output sequence. We used the word-level attention mechanism with sigmoid activation function which helps the model to focus on words which matter most for the meaning of the sentence. This attention is applied to the output of previous Bidirectional CuDNNLSTM Layer and returns an attention-weighted combination of the input text.

Convolution Operation The convolutional operation which is generally used in computer vision is also useful for feature extraction in text sequences. Convolution operation consists of Convolution layer, ReLU activation, and Pooling layers. The output features of attention layer is concatenated with the extracted features from the text and is provided as input to convolution layer. By varying the kernel size of the convolution filter and then concatenating their outputs, the model can detect a pattern of multiple sizes which are equal to kernel size. We here chose the kernel size of 2, 3, and 4, and for each kernel size passed through max and average pooling operation to concatenate finally to form a single feature vector. This output is then passed as input to fully connected dense layers.

Fully Connected Dense Layers The output of convolution operation is then passed to fully connected dense layers with ReLU as the activation function. We also use dropout after dense layers to prevent overfitting of the model. We use two dense layers one after the other along with dropout, and the output is then passed to the final representation layer.

Representation Layer The output of fully connected dense layers is passed to this final output layer which consists of softmax activation. This output layer decides the probability of given text being sarcastic or non-sarcastic dependent on given input and outputs that as 0 (non-sarcastic) or 1 (sarcastic).

4 Results and Analysis

We performed all the experiments on Nvidia Tesla K80 GPU in Google Colab online environment. We experimented using a wide range of hyperparameters and mentioned that gave an optimal result in the Table 3. Experiments included embedding vectors of size 50, 100, 200, and 300 as well as different bidirectional LSTM and convolution layer units of 64, 128, 256, and 512. We also experimented with various dropout rates in range 0.1–0.5 to avoid overfitting of the network. The results for both the news headlines dataset and tweets dataset are summarised in the Table 4.

We compared our model with other baseline deep learning models, namely LSTM, Bi-direction LSTM without Attention, Convolutional Neural Network (CNN) by training and testing them on both the datasets as for the proposed model and Table 5 clearly shows that proposed model outperforms them. All the hyperparameters were kept constant in all the experiments and Glove embeddings of 300 dimensions used. For both the datasets, the proposed model performed best in terms of accuracy while CNN produces the least accuracy.

From the results, we can also analyze that due to the noise in the data collected from the Twitter compared to the news headline dataset, it achieves less accuracy in the detection of sarcasm for all the models. This accuracy can be increased by further pre-processing the text from social media so that noise can be decreased.

Table 3 Hyperparameters Values

Hyperparameter	Value
Dimension of GloVe embedding vectors	300
Bidirectional LSTM units	128
Bidirectional CuDNNLSTM units	128
Convolution filters	256
Convolution kernel sizes	2, 3 and 4
Learning rate	2e–2
Batch Size	64
Loss	Binary crossentropy
Dropout rate	Word Embedding: 0.3 LSTM layer: 0.5 Dense layers: 0.2
Activation function of convolution and dense layers	ReLU

Table 4 Performance of proposed model

Parameter	News headlines dataset	Tweets dataset
Accuracy	89.19	87.87
F-Measure	89.04	87.59
Recall	89.20	87.77
Precision	89.19	87.97

Table 5 Accuracy comparison

Model	News headlines dataset	Tweets dataset
CNN	84.33	79.95
LSTM	86.05	82.55
Bidirectional LSTM without attention	87.40	83.72
Proposed model	89.19	87.87

5 Conclusion

Sarcasm is a specific type of sentiment, and it is essential to detect it automatically so that the sentiment of the text can be analyzed correctly. In this paper, we propose a new model with a combination of text-based features and deep learning model based on word embeddings, bidirectional LSTM, attention mechanism, and convolutional neural nets to detect sarcasm in social media text on Twitter as well as news headlines. The proposed model shows significant improvement compared to other deep learning-based methods on both the datasets achieving an accuracy of around 88 percent. The automated detection of sarcasm in the long text and changing vocabulary are considerable problems for the research in this domain.

References

1. Sarcasm. American Heritage® Dictionary of the English Language, 5th edn. Houghton Mifflin Harcourt Publishing Company (2019)
2. Maynard, D., Greenwood, M.: Who cares about sarcastic tweets? investigating the impact of sarcasm on sentiment analysis. In: Proceedings of the Ninth International Conference on Language Resources and Evaluation (LREC-2014), pp. 4238–4243. Reykjavik, Iceland (2014). European Languages Resources Association (ELRA)
3. Riloff, E., Qadir, A., Surve, P., De Silva, L., Gilbert, N., Huang, R.: Sarcasm as contrast between a positive sentiment and negative situation. Proceedings of EMNLP, pp. 704–714 (2013)
4. Hajic, J., Tsujii, J. (eds.): COLING 2014. In: 25th International Conference on Computational Linguistics, Proceedings of the Conference: Technical Papers. Dublin, Ireland. ACL (2014)
5. Joshi, A., Sharma, V., Bhattacharyya, P.: Harnessing context incongruity for sarcasm detection. In: Proceedings of the 53rd Annual Meeting of the Association for Computational Linguistics

and the 7th International Joint Conference on Natural Language Processing, vol. 2: Short Papers, pp. 757–762 (2015)

6. Tsur, O., Davidov, D., Rappoport, A.: Icwsm—a great catchy name: Semi-supervised recognition of sarcastic sentences in online product reviews. In: Fourth International AAAI Conference on Weblogs and Social Media (2010)
7. Liebrecht, C., Kunneman, F., van den Bosch, A.: The perfect solution for detecting sarcasm in tweets #not. In: Proceedings of the 4th Workshop on Computational Approaches to Subjectivity, Sentiment and Social Media Analysis, pp. 29–37. Atlanta, Georgia (2013). Association for Computational Linguistics
8. Joshi, A., Tripathi, V., Patel, K., Bhattacharyya, P., Carman, M.: Are word embedding-based features useful for sarcasm detection? (2016). arXiv:1610.00883
9. Ghosh, A., Veale, T.: Fracking sarcasm using neural network. In: Proceedings of the 7th Workshop on Computational Approaches to Subjectivity, Sentiment and Social Media Analysis, pp. 161–169 (2016)
10. Poria, S., Cambria, E., Hazarika, D., Vij, P.: A deeper look into sarcastic tweets using deep convolutional neural networks. CoRR (2016). ArXiv:abs/1610.08815
11. Ghosh, A., Veale, T.: Magnets for sarcasm: Making sarcasm detection timely, contextual and very personal. In: Proceedings of the 2017 Conference on Empirical Methods in Natural Language Processing, pp. 482–491. Copenhagen, Denmark (2017). Association for Computational Linguistics
12. Misra, R., Arora, P.: Sarcasm detection using hybrid neural network (2019). arXiv:1908.07414
13. Pennington, J., Socher, R., Manning, C.D.: Glove: Global vectors for word representation. In: Empirical Methods in Natural Language Processing (EMNLP), pp. 1532–1543 (2014)
14. Graves, A., Jaitly, N., Mohamed, A.: Hybrid speech recognition with deep bidirectional lstm. In: 2013 IEEE Workshop on Automatic Speech Recognition and Understanding, pp. 273–278 (2013)
15. Chetlur, S., Woolley, C., Vandermersch, P., Cohen, J., Tran, J., Catanzaro, B., Shelhamer, E.: Cudnn: efficient primitives for deep learning. CoRR (2014). ArXiv:abs/1410.0759
16. Luong, M.-T., Pham, H., Manning, C.D.: Effective approaches to attention-based neural machine translation. CoRR (2015). ArXiv:abs/1508.04025

Sentiment Analysis Using Twitter

Ujjayanta Bhaumik and Dharmveer Kumar Yadav

Abstract Analysing the sentiment from the tweets of a user can give general insights into how the person is thinking. This paper analyses different groups of people like graduate students, politicians and doctors and aims to study the behavioural patterns. A method was designed to extract sentiment from tweets obtained using Twitter API. The VADER sentiment analysis library was used to gather median sentiment scores for different groups of people. The results show that politicians have a similar propensity for positive and negative tweets. Comedians' tweets have the best positive median score, while graduate students' tweets have the least positive median score among the tested groups. This paper extensively compares tweets of different groups of people and provides a novel analysis of general sentiment trends.

Keywords Sentiment analysis · VADER · Twitter

1 Introduction

Sentiment analysis is a machine learning approach to gather knowledge about the emotional state of an author from their authored piece of text. This paper utilizes the VADER (Valence Aware Dictionary and sentiment Reasoner) sentiment analysis library to analyse the sentiment of different groups of people, namely politicians, doctors, comedians, motivational speakers and graduate students. The VADER library was specially developed for social media analysis. It is a simple rule-based model that is used to analyse typical social content using qualitative and quantitative methods. Sentiment analysis is used in various fields like marketing research, customer psychology studies, economics and business strategy development.

U. Bhaumik (✉) · D. K. Yadav
Katihar Engineering College, University College London (UK), Katihar, India
e-mail: ujjayanta.bhaumik.18@ucl.ac.uk

D. K. Yadav
e-mail: kumar.dharmveer@gmail.com

J. K. Mandal et al. (eds.), *Computational Intelligence and Machine Learning*, Advances in Intelligent Systems and Computing 1276,
https://doi.org/10.1007/978-981-15-8610-1_7

Twitter is a popular social media that focuses mainly on current world affairs and also allows users to be connected by allowing them to follow each other and also retweet other's tweets [1]. It is a valuable source of data as millions of users use the platform every day. The tweets are generally publicly available can be downloaded using Twitter's application programming interface. As a result, this makes it possible to study people's opinions, holistically and on the individual level too [1]. The social Twitter Chirp developer conference in April 2010 [2] provided some statistics about Twitter and the platform's user engagement.

According to Twitter, 3 lakh new users were added to the site every day and over 600 million queries populated its search engine, and above 3 billion requests per day were carried out based on the results of Twitter API. Nearly 40% of the users used their mobile to tweet. As of September 2019, Twitter had 330 million registered users. The past two decades have seen an enormous uprising in the use of social media platforms with the engagement of people from all corners of the world. This growth eventually attracted big companies and they started focusing on social media like Twitter. Because of the copious amount of data that can be mined, a lot can be learnt about the thinking process of the users. Companies like Twitratr, Tweetfeel and Social Mention use sentiment analysis on Twitter for their products and services. A lot of interesting sentiments come out from the daily tweets of users across the world. Online reviews and news articles are rich sources of information, also online messages in gaming platforms and other media constraints can be studied in detail. Different properties related to automatic POS tags or part of speech analysis and sentiment dictionaries created with the help of real human annotations are useful for sentiment analysis in various domains, including Twitter. In this paper, we begin to investigate this question regarding Twitter. In this paper, we explore one method for building such data using API and public twitter accounts to identify different sentiments in tweets like positive, negative and neutral tweets to use for studying different groups of people. Twitter is a really popular social media with over 300 million accounts. This makes it a gold mine of people's opinions for sentiment analysis [2]. For each tweet, one can determine the sentiment it portrays, whether the tweet is it positive, negative or neutral.

2 Literature Survey

Analysis of sentiment is considered as natural language processing work at many levels of granularity. Starting with a classification assignment at the document level [3, 4], it was addressed at the sentence level [5] and more lately at the sentence level [6, 7]. As tweet sentiment analysis has gained popularity in recent years, Naive Bayes classifier and page rank technique have been used to calculate user-generated sentiments [8]. Sentiment classification is done using unsupervised and supervised algorithms for unstructured data like tweets and SMS [9]. Many works have been done to analyse the tweet sentiments using machine learning [14, 15], the research work is proposed based on machine learning technique using Naive Bayesian technique,

SVM and entropy-based [16–18] to classify the review. In 2010, the US election dataset [10] was collected to determine and rank individuals politically on the basis of political discussion and data available on Twitter timelines. They divided the analysis into two wide classifications user level and network level for politically classifying individuals. Researchers have also started to explore different methods of training data automatically. In order to define their training data, several researchers work based on emoticons [11, 12]. Use current Twitter sentiment sites [20] to collect training data [13] also use hashtags to create training data, but their experiments are limited to the classification of sentiments/non-sentiments.

The Twitter Application Programming Interface (API) presently provides a Streaming API and two separate REST APIs. Users can access tweets in sampled and filtered form in real time via the Streaming API [16]. The API is based on HTTP and demands from GET, POST and DELETE can be used to access the information. Individual posts on Twitter represent a user's "status". With the help of Streaming API, user may access public status including responses in almost real time.

Twitter sentiment analysis has been a widely researched topic and can be traced back to a phrase analysis of sentiments. Phrase-level processing of sentiments was discussed by Wilson et al. [6] which was really innovative in categorizing phrases based on positive and negative behaviour. The paper pointed out the fact that some phrases were objective and used for sentiment analysis earlier. Those phrases led to the poor categorization of the objective phrases. It also found that if a simple classifier that assumes that the word's contextual polarity equals its prior polarity, a result of nearly 48% is obtained. Pak et al. designed a method for automatic collection of twitter data for building a large corpus using Tree Tagger in 2010 [11].

2.1 *Twitter Sentimental Analysis*

Social networks are an active source for sentiment analysis study as people are always attached to social media like Facebook and Twitter, discussing various topics, promoting business and so on. The sentiment-aware systems are used in business studies, social science analysis and making accurate predictions regarding human behaviour. Several abbreviations are used by users in their social media life as often people want to be quick and thus these terms have become a part of the normal social media vocabulary, like FB for Facebook and OMG for oh my god. Since social networks, especially Twitter, contains individual tweets that are small and also many abbreviations which might be difficult to interpret by general Natural Language processing, the VADER library has been used for analysis [21–23]. The Twitter API is used to collect tweets to create a dataset and VADER is used for analysis shown in Fig. 1.

We have used the following short messages in our approach.

Emoticons: Emoticons are predominantly used as shortcuts for emotions and.

VADER takes emoticons into account also.

Hashtags: Hashtags are used for organizing content on social media.

Fig. 1 Methodology describing the generation of median sentiment score

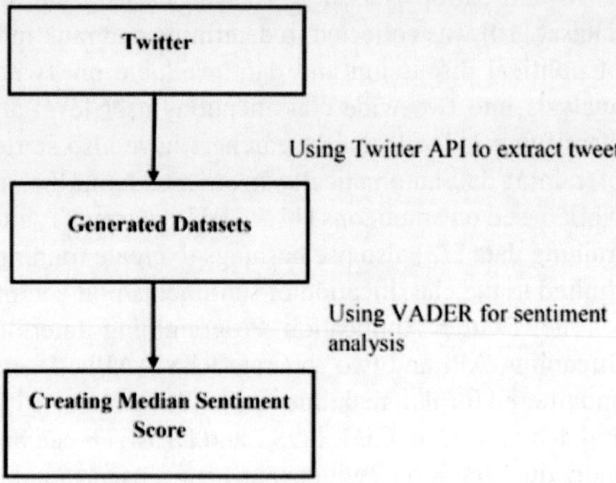

Normalize and correct spelling: This is used so that VADER results are not affected by changes in spelling.

Twitter API allows users to download tweets officially from a user account and the tweets can be saved in a suitable file format like csv. Using the API, tweets from different users across the five categories: graduate students, doctors, politicians, comedians and video gamers were downloaded and concatenated separately into datasets for different groups. After that, VADER was used for generating sentiment score for individual tweets and the median for each group was calculated using the individual tweet sentiment scores for each group.

The VADER sentiment analysis model working can be explained using the following steps:

- Seeing lexical features which are approved by humans and features that are already established
- Adding new features that cover the plethora of sentiments used in social media, for instance, the emojis, different Internet slangs and abbreviations
- Using an approach called crowd wisdom to estimate emotional value based on more than nine thousand lexical features
- Evaluating the effect of common grammar and syntactical rules on the sentimental quality of the text
- Applying coding analysis and results to identify general patterns in the text and identifying sentiment
- Everything is done with respect to a benchmark human sentiment gold standard emotional dictionary or sentiment lexicon.

3 Results and Discussion

A group of ten graduate students, ten politicians, ten doctors, ten comedians and ten video gamers who use Twitter were selected randomly. For each individual in a group, a corpus of 3200 tweets were created. The following figures show the average sentiment analysis results for persons from each group. The zero mark displays the tweets of a user where the sentiment is neutral, while less than zero indicates negative sentiment and more than zero indicates positive sentiment.

Fig. 2 shows the sentimental trends and changes based on the emotional nature of the tweets. The figure on the left represents the sentiment analysis results for video gamers. On average, video gamers tweets are mostly on the neutral and positive side, which is expected as gaming is related to the adrenaline rush and people and mostly excited and happy while playing games. The right figure shows a figure representing the sentiment analysis results for graduate students. Interestingly, graduate students have the least median sentiment score among all the groups.

The Fig. 3 shows tweets from the group comprised of doctors, politicians and comedians had high sentimental scores as compared to graduate students and video gamers. The left figure represents the sentiment analysis results for doctors. The right figure shows the sentiment analysis results for politicians and comedians. The graph with the two higher peaks represents the comedian sentiment. The politicians are

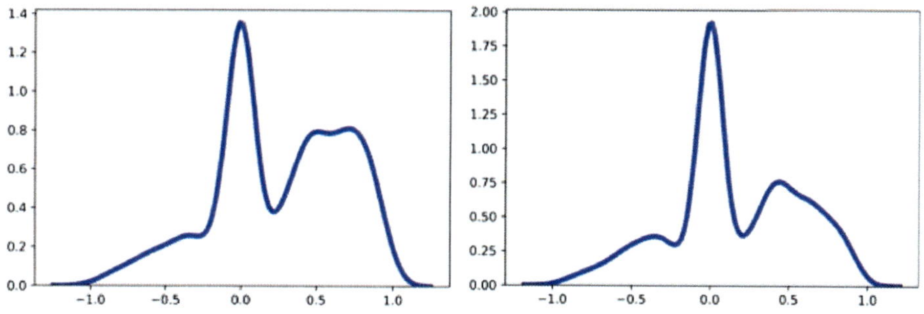

Fig. 2 Sentiment analysis results for video gamers and graduate students

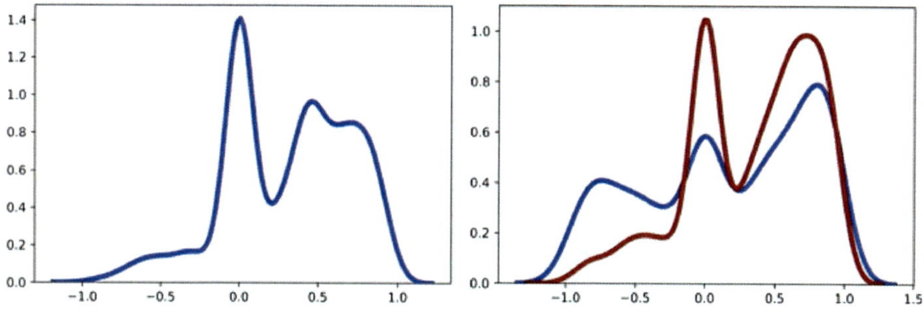

Fig. 3 Sentiment analysis results for doctors, comedians and politicians

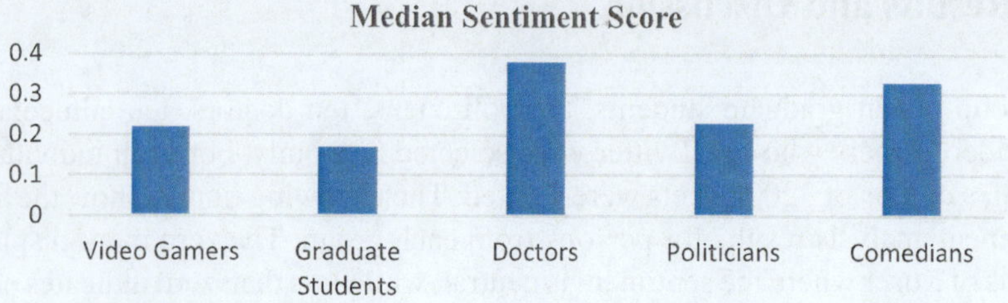

Fig. 4 Median sentiment score of students, politicians, doctors, comedians and video gamers

equivocal on an average with almost comparable amounts of positive and negative tweets, while most of the tweets by comedians are positive as expected. Figure 4 shows the Median sentiment score of students, politicians, doctors, comedians and video gamers. A similar sentiment analysis using Facebook data from students' comments in Bahagian Pengurusan Kewangan Pelajar UiTM Facebook page showed that the sentiments were mostly negative or neutral. In this work too, the students had the lowest median sentimental score. Also, VADER performs much better as compared to other similar sentiment analysers. The median positive sentiment was also calculated for each group separately. It is the median of the compound sentiment score for a tweet which shows how positive or negative a tweet is, for all the tweets in a particular group. The median sentiment score for each group is listed below:

The doctors had the highest median sentiment score of 0.38, while graduate students displayed the lowest median sentiment score of 0.17. The dataset was created by using Twitter API. Using the API, tweets were downloaded from Twitter and saved into a CSV file, from which sentiment analysis was done.

4 Conclusion

Sentiment analysis is a great tool to measure the emotional weight of a text and gives a good idea about the opinions expressed. The work focused on tweets by different groups of individuals. Social media is a great place for the study of emotions given their popularity in the last two decades. The study showed that graduate students had the least positive median sentiment score for their tweets, while doctors and comedians scored really high with their tweets. A chronological and sentimental evaluation approach to tweets can give insights about emotional statistics in an individual. The results obtained show the emotional differences based on a tweet database among the different groups.

5 Future Work

The study provides an innovative path for future research in sentiment analysis and for future studies, other social media like Facebook and Reddit can be considered as well. Different social media serve different purposes as well and thus comparing the results across different social media could be another possibility.

References

1. Schonfeld, E.: Mining the thought stream. TechCrunch Weblog Article (2009)
2. Neethu, M.S., Rajasree, R.: Sentiment analysis in twitter using machine learning techniques. In: 2013 Fourth International Conference on Computing, Communications and Networking Technologies, pp. 1–5. IEEE (2013)
3. Turney, P.D.: Thumbs up or thumbs down?: semantic orientation applied to unsupervised classification of reviews. In: Proceedings of the 40th Annual Meeting on Association for Computational Linguistics, pp. 417–424 (2002)
4. Pang, B., Lee, L.: A sentimental education: Sentiment analysis using subjectivity summarization based on minimum cuts. In: Proceedings of the 42nd Annual Meeting on Association for Computational Linguistics, p. 271 (2004)
5. Hu, M., Liu, B.: Mining and summarizing customer reviews. In: Proceedings of the Tenth ACM SIGKDD International Conference on Knowledge Discovery and Data Mining, pp. 168–177. ACM (2004)
6. Wilson, T., Wiebe, J., Hoffmann, P.: Recognizing contextual polarity in phrase-level sentiment analysis. In: Proceedings of Human Language Technology Conference and Conference on Empirical Methods in Natural Language Processing (2005)
7. Apoorv, A., Fadi, B., Kathleen, M.: Contextual phraselevel polarity analysis using lexical affect scoring and syntactic n-grams. In: Proceedings of the 12th Conference of the European Chapter of the AC, pp. 2432 (2009)
8. Liu, B.: Sentiment analysis and opinion mining. Synth. Lect. Hum. Lang. Technol. **5**(1), 1–167 (2012)
9. Rout, J.K., Choo, K.K.R., Dash, A.K., Bakshi, S., Jena, S.K., Williams, K.L.: A model for sentiment and emotion analysis of unstructured social media text. Electron. Commer. Res. **18**(1), 181–199 (2018)
10. Conover, M.D., Gonalves, B., Ratkiewicz, J., Flammini, A., Menczer, F.: Predicting the political alignment of twitter users. In: 2011 IEEE Third International Conference on Privacy, Security, Risk and Trust and 2011 IEEE Third International Conference on Social Computing, pp. 192–199. IEEE (2011)
11. Pak, A., Paroubek, P.: Twitter as a corpus for sentiment analysis and opinion mining. LREc **10**, 1320–1326 (2010)
12. Bifet, A., Frank, E.: Sentiment knowledge discovery in twitter streaming data. In: International Conference on Discovery Science, pp. 1–15. Springer, Berlin (2010)
13. Davidov, D., Tsur, O., Rappoport, A.: Enhanced sentiment learning using twitter hashtags and smileys. In: Proceedings of the 23rd International Conference on Computational Linguistics: Posters, pp. 241–249. Association for Computational Linguistics (2010)
14. Sebastiani, F.: Machine learning in automated text categorization. ACM Comput. Surv. (CSUR) **34**(1), 1–47 (2002)
15. Pang, B., Lee, L., Vaithyanathan, S.: Thumbs up?: sentiment classification using machine learning techniques. In: Proceedings of the ACL-02 Conference on Empirical Methods in Natural Language Processing, vol. 10, pp. 79–86. Association for Computational Linguistics (2002)

16. Gautam, G., Yadav, D.: Sentiment analysis of twitter data using machine learning approaches and semantic analysis. In: 2014 Seventh International Conference on Contemporary Computing (IC3), pp. 437–442. IEEE (2014)
17. Joachims, T.: Text categorization with support vector machines: Learning with many relevant features. In: European Conference on Machine Learning, pp. 137–142. Springer, Berlin (1998)
18. Berger, A., Pietra, V.J.D., Pietra, S.A.D.: A maximum entropy approach to natural language processing. Comput. Linguist. **22**(1), 39–71 (1996)
19. Kim, S.M., Hovy, E.: Determining the sentiment of opinions. In: Proceedings of the 20th International Conference on Computational Linguistics (2004)
20. Barbosa, L., Feng, J.: Robust sentiment detection on twitter from biased and noisy data. In: Proceedings of the 23rd International Conference On Computational Linguistics: Posters, pp. 36–44. Association for Computational Linguistics (2010)
21. Elghazaly, T., Mahmoud, A., Hefny, H.A.: Political sentiment analysis using twitter data. In: Proceedings of the International Conference on Internet of things and Cloud Computing, pp. 1–5 (2016)
22. Wilson, T., Wiebe, J., Hoffmann, P.: Recognizing contextual polarity inphrase-level sentiment analysis. In: Proceedings of Human Language Technology Conference and Conference on Empirical Methods in Natural Language Processing (2005)
23. Hutto, C.J., Gilbert, E.: Vader: a parsimonious rule-based model for sentiment analysis of social media text. In: Eighth International Conference on Weblogs and Social Media (2014)

A Type-Specific Attention Model For Fine Grained Entity Type Classification

Atul Sahay, Kavi Arya, Smita Gholkar, and Imon Mukherjee

Abstract Different methods exist for Named Entity Type Classification (NETC), ranging from primitive rule-based learning to Deep Learning methods. This paper discusses notable parts of significant methods that exist in the domain. In the process of researching these methods, we observed certain drawbacks in the current Fine Grained Entity Type Classification (FETC) methods. Ignoring local contextual linguistic information and ignoring type membership information of these contextual entities are two main drawbacks we have focused upon. We propose an approach that overcomes both these drawbacks by employing a Bi-Directional Short Term Memory (Bi-LSTM) network with a layer of "type-specific" attention. Our approach achieved $\sim$2% increased improvement on F1 scores over and above the state-of-the-art work on the publicly available dataset FIGER (GOLD) with considerably lesser training time.

Keywords Named entity recognition and classification · Fine grained entity type classification · Bi-directional long short-term memory · Attention weighted averaging · Contextual linguistic representation · Adversarial learning

---

A. Sahay (✉) · K. Arya · S. Gholkar
Indian Institute of Technology, Bombay, Maharashtra, India
e-mail: atulsahay@cse.iitb.ac.in

K. Arya
e-mail: kavi@iitb.ac.in

S. Gholkar
e-mail: smitagh@cse.iitb.ac.in

I. Mukherjee
Indian Institute of Information Technology, Kalyani, West Bengal, India
e-mail: imon@iiitkalyani.ac.in

J. K. Mandal et al. (eds.), *Computational Intelligence and Machine Learning*, Advances in Intelligent Systems and Computing 1276,
https://doi.org/10.1007/978-981-15-8610-1_8

1 Introduction

Information extraction (IE) is the task of automatically extracting known or semi-known information from unstructured and/or semi-structured data. In the Information Extraction (IE) Domain, Named Entity Recognition (NER) is a Natural Language Processing (NLP) technique which focuses on mainly the location and labeling of the named entity. A named entity can be mentioned in a collection of documents as one of pre-defined categories of labels such as person names, organisation, locations, monetary values, sports expressions for Time etc; thereby annotating the text [13].

Let us consider an example of a text sentence as stated below:

$$\textbf{Adolf Hitler came to power in 1933.} \tag{1}$$

Producing annotation to the text will give us the following information:

$$\textbf{[Adolf Hitler]}_{person}\ \textbf{came to power in [1933]}_{time} \tag{2}$$

In the given example (2), a two-token name and a one-token time expression is classified into corresponding tags viz., Person and Time. By token here we refer to individual word combinations.

Named entities mentioned in contextual documents, restrict the task of NER to one or many alphanumeric strings, such as words or phrases, which when referred to (referent) performs fairly well. American philosopher and logician Kripke coined the NER term as rigid designators, because of the rigidity Named entities have in a mention [13]. Although in practice, NER models deal with scenarios where names and references of the mentioned entities are not philosophically rigid at all. For example lets consider the name "FORD Motor's" founded by Henry Ford, here "Ford" can actually refer to many other entity

People

- Ford and Mistress Ford, inked in William Shakespeare's *The Merry Wives of Windsor*

Organization

- Ford motors, established by Henry and Edsel.

Place

- Ford, Argyll

Clearly, it can be seen that rigid designators includes proper noun and referent while they exclude pronouns like "it" even though it judges on the basis of properties of the referent . $[Hindujas]_{organization}$ are the richest citizens in $[Uk]_{location}$. They have recently acquired an old war building in $[Uk]_{location}$. From the example stated above, if we take the consideration of referent properties then "They" will be flagged as an entity with the tag of a person or an organisation. This causes an ambiguity between the label of the named entity as a person or organisation.

This field of disambiguation in NER is known as **"Coreference Resolution".**

The general dissolution of full NER, conceptually and implementation wise, is done as two distinct problems: detection of names and labelling of the names into one of the many tags available. The detection task obviously deals with segmentation, as particularly defined named entities can span over a number of words excluding any word nesting, such that "Bank of Baroda" can be segmented as one entity; disregarding the fact that "Baroda" itself is a named entity. While labelling systematically deals with choosing most particular ontology to categorize the named entities.

For better correlation with the domain on which NER is modeled, strictness of named entities are loosened over the context, (e.g., 2001, June 19) is a rigid case of named entity where it is clearly mentioned that date happens to be 19th of June of 2001 while in "Ram works in June", it is not clear which year's June we are talking about. Because of unseen behaviour of NER task over a domain, NER task defined over one domain may not work at all in other domains. This is a problem in NLP, commonly known as brittleness of a model, where it means models can't be generalised further.

The main contribution of the paper are as follows:

1. A clean and brief transition of NER tasks
2. An improvement over the Shimaoka's state-of-the-artwork [12] Attentive Fine-Type Tagger on the FIGER GOLD dataset [10] is duly presented.

1.1 NETC

The term "named entity" was coined in [2, 4], by Lisa F. Rau, where the author used named entity to describe the task of extraction of company names from unstructured text, such a newspaper and company record base. This led to breakthrough work of extraction of structured units from the unstructured units in the following years by researchers [3].

Till 2003, no supervised learning methods were performed in NER tasks; even large search engines such as Yahoo were solely dependent on the simpler rule-based learning [see Algorithm 1]—where patterns are 3 tuples consisting of—1. prefix of the referent, 2. named entity itself and 3. suffix of the referent [see Eq. 3]. In rule-based learning methods, the pattern set,P is defined with seed examples.

$$P_i = \langle prefix, \text{named entity}, suffix \rangle \tag{3}$$

Since the year 2003 [1], IE domain has seen supervised learning through discriminative models such as Conditional Random Field (CRF) or statistical models; which was used to perform the task of entity extraction and its type classification. It is not the same with supervised learning based models and thus rule-based IE models became obsolete. Till that time rule-based learning was still in practice because no model could match human perception. CRF-based information extraction models used to

```
Result: Extraction of Rule Set R
P ⟵ P_1, ···, P_n;
while GenerateRule(P).isNotEmpty() do
    R_x = GenerateRule(P);
    if R_x matches max(P) then
        R ⟵ add(R_x);
        P ⟵ extract(R);
    else
        Reject R_x;
    end
end
```

Algorithm 1: Rule-Based Named Entity Extraction

dominate the NLP societies up till 2011 [5]; that is when the first deep learning model was employed for NERC.

1.2 FETC: Refining the Granularity

With refined granularity of entity type set, more latent information can be extracted from the document which can be useful for the purpose of many other IE tasks such as relation extraction. (Xiao Ling and Weld) [10] defined the fine grained tag set of 112 tags along with the multi-class multi-label classification model FIGER, where learning happens via typing a CRF based model for segmentation with a classifier, to label the extracted entity set, using the multi class multi label heuristics defined in Eq. 4.

$$\hat{y} = argmax_y w^T \cdot f(x, y)\ \hat{y} \epsilon T \tag{4}$$

Here T is the possible number of tags, x be feature inputs, f be the scoring function and w is the weight defined for the inference.

1.3 Recent Advances in FETC

The Deep Learning era began with the temporal Convolutional Neural Network(CNN) over a sequence of words of Collobert [5] . It then went on to the replace CNN with a Bi-LSTM architecture by Huang [6]. The most recent method by Chiu and Nicolas work introduced hierarchy in the character—level feature set using hand-engineered features where Bi-LSTM and CNN structure has been demonstrated [7].

The task of FETC are not limited only to entity chunking but go further to new problem sets of Relation Extraction, Coreference Resolution or Record Linkage or Pronoun Disambiguation, Named Entity Disambiguation, Named Entity Linkage. This problem set can further collate bigger problems of Question and Answering

System, Knowledge Base Construction or Implantation or Completion. The success of the search engine, GOOGLE is largely based on the effectiveness of NER task as stated in DIPRE [8]. The underline working structure of the DIPRE model is similar to the what has been discussed in the Algorithm 1. One of the finest work in the field of FETC is the use of adversarial learning. In machine learning, adversarial learning may be coined as the flow of learning based on adversarial examples. Adversarial networks employ two separate networks—generative and discriminative. The sole purpose of adversarial learning is to make training harder than testing so that a model can stay robust with test errors. Adversarial learning is like a 2-player min-max game, where one tries to generate (Generative Network) as hard as possible false data (noise data) and the other tries to counterfeit the false data (real data) (Discriminative Network). Each network tries its best to outperform the other at their game. When they balance each other out, we get our required model for the task of inference. DATnet [9] is one such novelty done for FETC in Chinese language.

The remaining paper is structured as follows: Sect. 2 provides brief discussions about the workflow of our proposed model over the Shimaoka's FETC model [12], Sect. 3 presents the experimental results of our proposed model that validate the feasibility and reliability of our work. Finally we conclude in Sect. 4.

2 Proposed Model

We formulated our model's flow as the FETC problem, where given an entity mention with its left and right context as $l_1, \ldots, l_c, m_1, \ldots, m_k, r_1, \ldots, r_c$; where c is the context span undertaken as the context locality and k is the mention entity span.

Given this formulation, we then go on compute the probability $y_i \in$ R for each of the fine grained tag 'T' types. We then outputted tag t whose y_t is greater than the threshold undertaken.

2.1 Embedding Space

The embedding acquisition [11] for FETC allows the sharing of information between the entities which are close in the embedding space if they frequently co-occur with each other. The token vectors are mapped to lower dimensional space—this relationship can be visualised from the proximity of the vectors in the embedding space.

For each of the input word vector v_i(either a mention or a context vector), we brought it to lower dimensional embedding space H:

$$p(x) : \mathbb{R}^L \rightarrow \mathbb{R}^H \tag{5}$$

Feature function p(x) denotes the mapping functions for converting vectors to the lower dimensional embedding space H.

2.2 Mention Vector Representation

Once each word vector which is brought down to lower dimensional embedding space, we formulated a mention representation as defined in equation :

$$v_m = \frac{1}{|m|} \sum_{|m|}^{1} u(m_i) \tag{6}$$

Where the mention vector v_m is given by averaging the word embeddings of the mention words $m_1, m_2, \ldots, m_{|m|}$ and $u(m_i)$ is the word embedding for the word m_i.

2.3 Context Vector Representation

Shimaoka's Approach: Attentive Neural Model Context entities present influential and latent characteristics of the mentioned entities. So to learn the context vector Shimaoka [12] employed the Attention weighted Bi-LSTM

Assuming the standard Notation:

$$Softmax(z_1, z_2, \ldots, z_m) = \frac{(\exp z_1, \ldots, \exp z_m)}{\sum_m^M \exp z_m} \tag{7}$$

$$(\overrightarrow{h_1}, \overleftarrow{h_1}), \ldots, (\overrightarrow{h_C}, \overleftarrow{h_C}) = BiLSTM(x_1, \ldots, x_C) \tag{8}$$

where the pair $(\overrightarrow{h_i}, \overleftarrow{h_i})$ represents the state vector in the forward and backward direction respectively. C = context word length and $\overrightarrow{h_i}, \overleftarrow{h_i} \epsilon R^{H\times 1}$.

In the paper, the context vector v_c is learnt using attention weighted contextual representation by passing state vectors pairs to the fully connected layer with D_a hidden units in a hidden layer, $e_i \epsilon \mathbb{R}^{D_a\times 1}$ and weight matrix $W_e \epsilon R^{D_a\times 2H}$. For each such hidden state vector e_i, a scalar unit a_i is learnt with the weight matrix $W_a \epsilon R^{1\times D_a}$. These scalar unit are then normalized to 1 and this whole formulation shown in Fig. 1 forms the attention mechanism over the context word vectors as defined in Eq. 13.

$$\mathbf{z_i}^{\mathbf{l}} = \begin{bmatrix} \overrightarrow{h_i}^l \\ \overleftarrow{h_i}^l \end{bmatrix} \tag{9}$$

$$e_i{}^l = \tanh\left(W_e \times \mathbf{z_i}^{\mathbf{l}}\right) \tag{10}$$

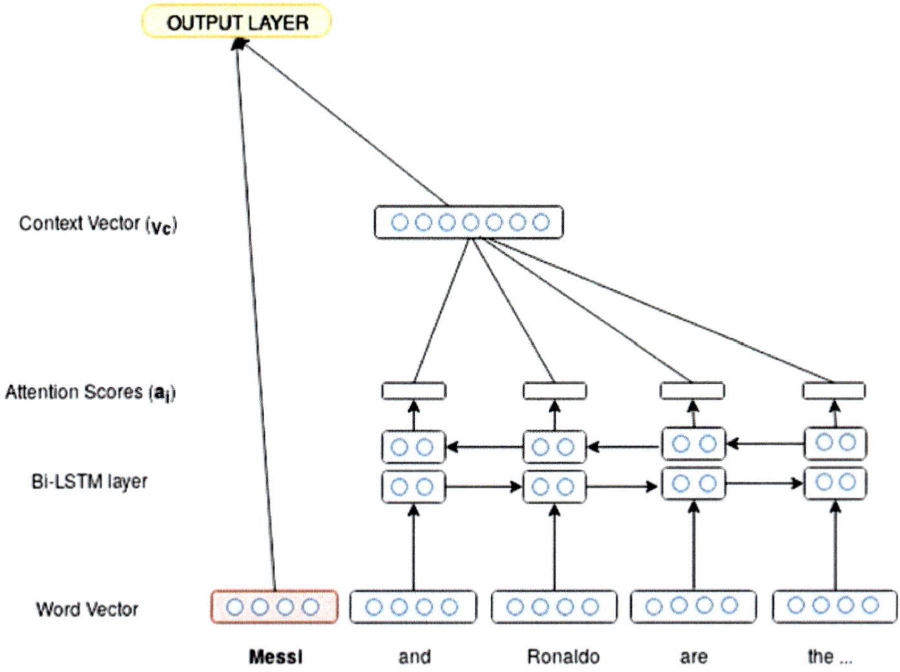

Fig. 1 Shimaoka's attention model for prediction of mentioned entity—Messi

$$\tilde{a_i}^l = \exp(W_a \times e_i{}^l) \quad (11)$$

$$a_i{}^l = \frac{\tilde{a_i}^l}{\sum_{i=1}^{C}(\tilde{a_i}^l + \tilde{a_i}^r)} \quad (12)$$

$$\mathbf{v_c} = \sum_{i=1}^{C}(a_i{}^l \dot{\mathbf{z_i}}^{\mathbf{l}} + a_i{}^r \dot{\mathbf{z_i}}^{\mathbf{r}}) \quad (13)$$

Our Approach: Type-Specific Attention Tag information of the local context in addition to the attention weighted contextual linguistic expression [shown in Eq. 13] can give much needed confidence to the model to predict the tag type of the mention entity.

In our model, we tried to first learn the type specific attention vectors V_t for each of the fine grained tag 'T' types and then a scalar unit vector B is learnt [Eq. 15] and then these scalar vector elements are normalized to 1. This whole formation provides the type-specific attention as shown in Fig. 2, which was then weighted with context words to form the context vector v_c. Assuming the $y \epsilon [1, \ldots, T]$, we replaced the attention weight $W_a \epsilon R^{1\times D_a}$ to $W_a \epsilon R^{T\times D_a}$

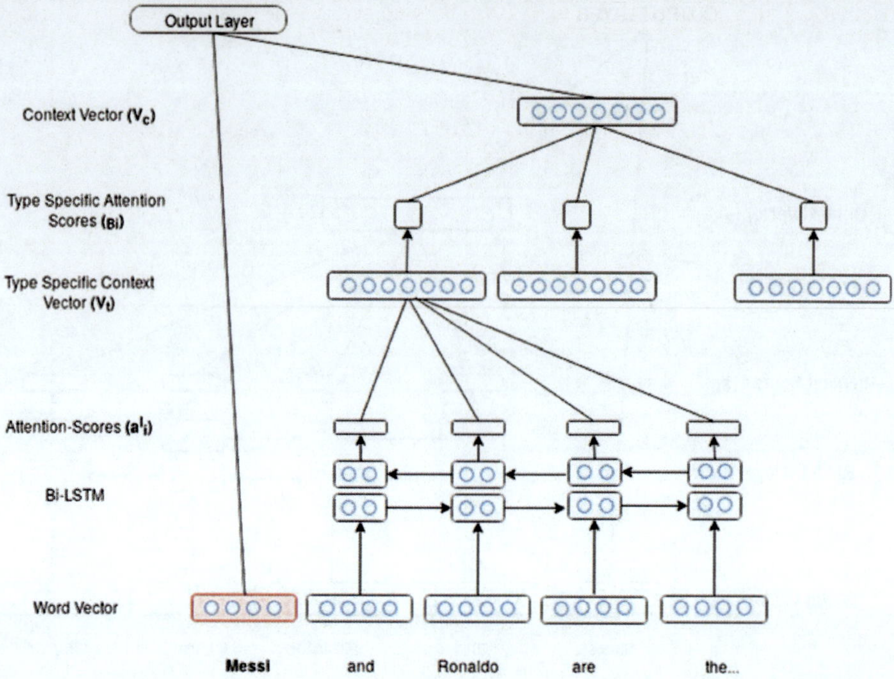

Fig. 2 An illustration of a type-specific attention layer over Bi-LSTM mode for prediction of mentioned entity—Messi

$$\forall t \epsilon [T] : V_t = \sum_{i=1}^{C} (a_i{}^l \odot \mathbf{z_i}^{\mathbf{l}} + a_i{}^r \odot \mathbf{z_i}^{\mathbf{r}});\ V \epsilon R^{T \times 2H} \tag{14}$$

$$\tilde{B} = diag(W_e \times V);\ \tilde{B} \epsilon R^{T \times 1} \tag{15}$$

$$B = Softmax(\tilde{B});\ B = (b_1, \ldots, b_T) \tag{16}$$

$$v_c = \sum_{t=1}^{T} (b_t \odot V_t) \tag{17}$$

2.4 Inference and Loss

The logistic regression layer learns the probability distribution $y_1, \ldots, y_T$ for 'T' tag classes by making use of embeddings of the words of mention and their context: $v_m \epsilon R^{D_m \times 1}$ and $v_c \epsilon R^{D_c \times 1}$, where $W_y \epsilon R^{T \times (D_c + D_m)}$ is the weight matrix.

$$y = \frac{1}{1 + \exp\left(- W_y \begin{bmatrix} v_m \\ v_c \end{bmatrix} \right)} \tag{18}$$

Following cross entropy loss function is employed in our learning:

$$L(y, \hat{y}) = \sum_{t=1}^{T} -y_t log(\hat{y}_t) - (1 - y_t) log(1 - \hat{y}_t) \tag{19}$$

where $\hat{y}_t$ is the predicted probability for the tag type 't' and $y \epsilon \{0, 1\}^{T \times 1}$ is the true tag type.

3 Evaluation and Results

For comparative analysis, we adapted the same evaluation approach as discussed in Shimaoka [12].

3.1 Dataset

A training size of 2,000,000, dev size of 10,000 and test size of 563 is used from the FIGER (GOLD) dataset [10]. A total of 112 overlapping fine grained type tags are present in FIGER. Training and development were curated from Wikipedia, whereas the test set containing the manual annotation was curated from news paper articles.

3.2 Evaluation Criteria

We used the same evaluation metric viz., strict, loose macro, loose micro as used by Shimaoka in his study.

Suppose 'N' is the size of the test set and P_i is the ground truth label and $\hat{P}_i$ is the predicted label for the i-th instance then the metrics are defined as follows:

- Strict

$$\text{Precision} = \text{Recall} = \frac{1}{N} \sum_{N}^{i=1} 1 \left[P_i = \hat{P}_i \right] \tag{20}$$

– loose macro

$$\text{Precision} = \frac{1}{N}\sum_{i=1}^{N}\frac{\left|\hat{P}_i \cap P_i\right|}{\left|\hat{P}_i\right|} \tag{21}$$

$$\text{Recall} = \frac{1}{N}\sum_{i=1}^{N}\frac{\left|\hat{P}_i \cap P_i\right|}{|P_i|} \tag{22}$$

– loose micro

$$\text{Precision} = \frac{\sum_{i=1}^{N}\left|\hat{P}_i \cap P_i\right|}{\sum_{i=1}^{N}\left|\hat{P}_i\right|} \tag{23}$$

$$\text{Recall} = \frac{\sum_{i=1}^{N}\left|\hat{P}_i \cap P_i\right|}{\sum_{i=1}^{N}|P_i|} \tag{24}$$

3.3 Results

From the comparative study between the results reported by Shimaoka in Table 1 and results obtained from our proposed model in Table 2, we got a performance improvement of $\sim 2\%$ in each scoring types discussed in Sect. 3.2, though we ran our experiments for smaller train duration and didn't fine tuning any hyper-parameters.

Table 1 Performance on FIGER (GOLD) [10] dataset by Shimaoka's attention model [12]

Scoring-type	Precison	Recall	F1-score
Strict	0.5595	0.5595	0.5595
Loose (Macro)	0.7417	0.7959	0.7679
Loose (Micro)	0.6881	0.7884	0.7348

Table 2 Performance on FIGER (GOLD) dataset by type specific attention model

Scoring-type	Precison	Recall	F1-score
Strict	0.5790	0.5790	0.5790
Loose (Macro)	0.7693	0.8021	0.7853
Loose (Micro)	0.7082	0.7923	0.7479

4 Conclusion

Our paper focuses on Fine-grained entity type classification methods, their drawbacks and overcoming them with recent development in its domain. As seen from the tabulated results, our proposed approach achieves a performance improvement of average 2% on every scoring type for the FIGER (GOLD) Table 1 dataset in comparison to what Shimaoka [12] achieved in his approach Table 2.

In future we propose fine tuning hyper-parameters such as: number of type tags, the specific granularity of a tag, etc. Longer training times to validate further improvement of F1 score accuracy may be looked into. Also, in real life scenarios, the contextual tag sets are not balanced in nature which contribute to erroneous or biased results. This can be overcome by using the specific granularity of a tag set i.e., using person/athlete and organisation/sports_team ; rather than the whole set of finer granular tags.

References

1. Finkel, J.R., Grenager, T., Manning, C.: Incorporating non-local information into information extraction systems by Gibbs sampling. In: Proceedings of the 43rd Annual Meeting on Association for Computational Linguistics (ACL 2005), pp. 363—370. Association for Computational Linguistics, Stroudsburg, PA, USA (2005). https://doi.org/10.3115/1219840.1219885
2. Nadeau, D., Sekine, S.: National Research Council Canada/New York University A survey of named entity recognition and classification. https://doi.org/10.1075/li.30.1.03nad
3. Sundheim, B.M.: Overview of results of the MUC-6 evaluation. In: Proceedings of the 6th Conference on Message Understanding (MUC6 1995), pp. 13–31. Association for Computational Linguistics, Stroudsburg, PA, USA (1995). https://doi.org/10.3115/1072399.1072402
4. Rau, L.F.: Extracting company names from text. In: Proceedings of Conference on Artificial Intelligence Applications of IEEE (1999)
5. Collobert, C., Weston, J., Bottou, L., Karlen, M., Kavukcuoglu, K., Kuksa, P.: Natural language processing (almost) from scratch.J. Mach. Learn. Res. **12**, 2493–2537 (2011)
6. Huang, Z., Xu, W., Yu, K.: Bidirectional lstm-crf models for sequence tagging (2015). arXiv:1508.01991
7. Chiu, J.P.C., Nichols, E.: Named entity recognition with bidirectional lstm-cnns. Trans. Assoc. Comput. Linguist. **4**, 357–370 (2016)
8. Brin, S.: Extracting patterns and relations from the World Wide Web. In: Proceedings of the 1998 International Workshop on the Web and Databases (WebDB 1998) (1998)
9. Zhou, J.T., Zhang, H., Jin, D., Zhu, H., Goh, R.S.M., Kwok, K.: DATNet: dual adversarial transfer for low-resource named entity recognition (modified: 21 Dec 2018). In: ICLR 2019 Conference Blind Submission
10. Ling, X., Weld, D.S.: Fine-grained entity recognition. In: Twenty-Sixth AAAI Conference on Artificial Intelligence (2012)
11. Yogatama, D., Gillick, D., Lazic, N.: Embedding methods for fine grained entity type classification. In: Proceedings of the 53rd Annual Meeting of the Association for Computational Linguistics and the 7th International Joint Conference on Natural Language Processing, vol. 2: Short Papers), vol. 2 (2015)
12. Shimaoka, S. et al.: An attentive neural architecture for fine-grained entity type classification (2016). arXiv:1604.05525
13. Wikipedia (2019). Named Entity Recognition Retrieved from https://en.wikipedia.org/wiki/Named-entity_recognition

A Two-Phase Approach Using LDA for Effective Domain-Specific Tweets Conveying Sentiments

Pradnya Bhagat and Jyoti D. Pawar

Abstract Twitter is a free social networking platform where people can post and interact with short messages known as "Tweets". The freedom of being able to reach out to the world in a fraction of seconds has made Twitter an effective medium for the general public to express their opinion on a global scale. Since Tweets have the potential to make a global impact, companies too have started using the service to reach out to their customers. Moreover, in spite of this service being immensely effective, it is found challenging by many users to express their views through a Tweet due to the restriction imposed of minimum 280 characters. The proposed work is aimed at helping people compose better quality Tweets belonging to a specific domain in the restricted character limit. The system is designed to mine important features/topics about a domain using Latent Dirichlet Allocation (LDA) algorithm and to compute the polarity of the sentiment words associated with them with respect to the domain using a two-phase approach on an Amazon review corpus. The discovered topics/features and sentiments are recommended as suggestions to Twitter users while composing new Tweets. The paper describes and presents initial results of the system on cell phones and related accessories domain.

Keywords Social networking · Twitter · E-commerce · Product review Tweets · Recommendations · Topics · Sentiment words · Latent Dirichlet Allocation

1 Introduction

Social media has started playing a pivotal role in our day-to-day life. Facebook, Twitter, Instagram, LinkedIn are examples of some of the networks developed to cater to the rising and constantly changing needs of the society [1]. Twitter [15] is

P. Bhagat (✉) · J. D. Pawar
Goa Business School, Taleigao, Goa, India
e-mail: dcst.pradanya@unigoa.ac.in

J. D. Pawar
e-mail: jdp@unigoa.ac.in

J. K. Mandal et al. (eds.), *Computational Intelligence and Machine Learning*, Advances in Intelligent Systems and Computing 1276,
https://doi.org/10.1007/978-981-15-8610-1_9

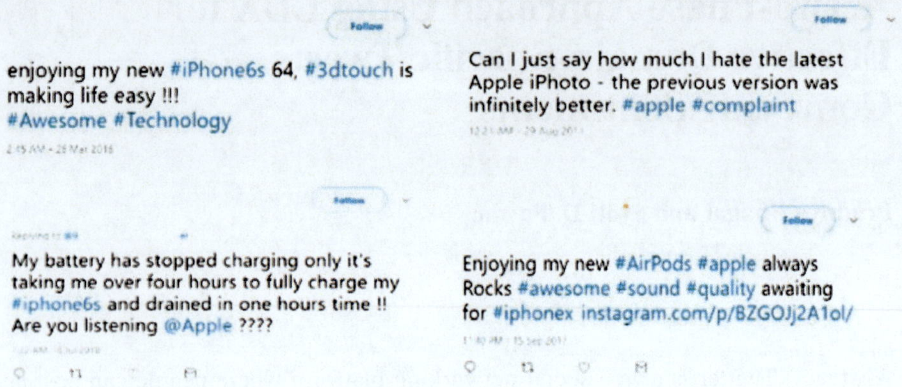

Fig. 1 Examples of Tweets about products [15]

a platform where users can interact with each other with the help of short messages knows as "Tweets". It is basically a micro-blogging platform where the message length is restricted to 280 characters. Since it is a free service with a global reach, people all over the world have started taking advantage of the same. Of late, we have seen an increase in the trend by people to review various products/services used by them on Twitter [5]. Twitter is also emerging as a platform to lodge complaints about goods or services having suffered the inconvenience. Being a broadcasting medium, the Tweeted message has the potential to instantly reach thousands of people creating global impact. Many other users reading this Tweet can base their purchase decisions on the read Tweet. This forces the companies to give immediate attention and address the issues faced by customers. Hence, Twitter is emerging as an effective medium for the public to make their problems heard and addressed. Figure 1 shows examples of some Tweets written by users addressed to the companies.

Although being such a powerful platform, we see that the use of Twitter is mostly restricted to specific sections of society. It has still not become successful in reaching to general masses. One of the reasons can be attributed to the 280 character limit for each Tweet, which acts as a great challenge to express one's opinions in brief. Especially to people whom English may be a foreign language, expressing one's views in such a constrained manner may not be always possible. The proposed system is designed with an aim to aid people; compose better quality Tweets related to products by recommending them with various product features and domain-specific sentiment words while composing Tweets. The system is unique from the fact that recommendations to help generate Tweets are sourced from amazon.com [2] which is an e-commerce website. Since there is no restriction on the length of the reviews on Amazon, it is found that Amazon reviews are more informative and descriptive than Tweets. Hence, reviews are used to generate recommendations. The discovery of topics from reviews is done using Latent Dirichlet allocation (LDA) [4] algorithm. The paper also proposes a method to find the polarity of sentiment words with respect to a particular domain since the sentiment of a word can largely depend on the context used. Since we are using the e-commerce domain to scrap reviews in order to develop recommendations for writing on Twitter which is a social networking platform, we can say that the proposed system is also a step towards cross-domain recommender systems.

2 Literature Survey

A significant amount of literature is being continuously contributed in the fields of social networking and e-commerce to keep up with the ever-evolving areas. The research depends heavily on Natural Language Processing and Machine Learning algorithms.

Dong et al. [7] describes a tool called Reviewer's Assistant that can be added as a browser plug-in to work with e-commerce websites like Amazon to help users write better quality reviews. The tool mines important topics from already published reviews and presents them as suggestions to new writers. Blei et al. [4] presents the Latent Dirichlet Allocation (LDA) algorithm, a generative probabilistic model for a collection of discrete data such as document collections. Dong et al. [6] describes a novel unsupervised method to extract topics automatically from a set of reviews. The work uses LDA algorithm and presents a major improvement on previous methods which required manual intervention. ki Leung et al. [10] proposes a probabilistic rating inference model for mining user preferences using existing linguistic processing techniques from a set of reviews and mapping the preferences on a rating scale. The method allows semantically similar words to have different sentiment orientations and hence tries to address the limitations of existing techniques. Zhang and Liu [16] presents a method to identify product nouns that imply opinions based on statistical tests.

3 Methodology

The work addresses the problem of assisting users to write better quality reviews using a two-phase approach:

1. Recommending topics/features in which the user may wish to express his/her opinions.
2. Suggesting appropriate opinion words to convey sentiments about the stated topics/features.

3.1 Recommending Product Features

Generally, many users are unaware of the technical terminology related to products. As a result, they are unable to use the correct terminology to explain the problem technically. Moreover, on a platform like Twitter, using elaborate sentences to explain one's problem is not feasible due to the character limit imposed. The proposed method uses LDA algorithm to address the issue. The approach adopted is as follows:

Reviews are broken down into sentences since most of the times a single sentence describes a single topic. The extracted sentences are Parts-of-Speech (POS) Tagged [13, 14] to identify the various parts of speech in the review. Since, most of the times, nouns are the parts of speech that convey the topic/features of the products, only nouns need to be retained from the entire review text. The challenge lies in identifying feature nouns from non-feature nouns.

In general, it is seen that the feature nouns occur in close proximity to sentiment words since the users are interested in expressing their opinions about the same. This is not the case with non-feature nouns. For example:

My friend advised me to buy this awesome mobile because it has this stunning look and attractive features.

POS Tagging of the above sentence would give us:

My_PRP friend_NN advised_VBD me_PRP to_TO buy_VB this_DT awesome_JJ phone_NN because_IN it_PRP has_VBZ this_DT stunning_JJ look_NN and_CC attractive_JJ features_NNS.

The nouns occurring in the above text are *friend*, *phone*, *look* and *features*. Out of these, the nouns we would be interested are *phone*, *look* and *features* since they belong to features/topics about cell phones. Further, as can be seen, nouns *phone*, *look* and *features* have some adjective associated with them since the user wants to express his opinions about the features, but noun *friend* does not have any adjective associated with it. This observation is utilized to differentiate feature nouns from non-feature nouns.

Next, the sentence position of the feature nouns is retained and the pre-processed file is given to the LDA algorithm. LDA is a statistical algorithm used to automatically identify topics across documents [4]. The algorithm follows the bag-of-words approach and considers documents as a set of topics that are made up of words with certain probabilities. As per the working of the algorithm, the words forming similar topics get grouped together. Whenever a person starts to compose a Tweet on any topic, many other words related to the same topic get displayed to the user. This can act as a valuable aid to the users to get correct technical words to express information effectively in the specified character limit.

3.2 Suggesting Appropriate Opinion Words

The second important part of the proposed framework deals with suggesting sentiment words or opinions about the topics to be addressed. Most of the works in the literature focus on selecting the adjectives based on POS Tagging and comparing them against a sentiment lexicon to find the polarity (positive/negative/neutral) of the opinions.

Although in the majority of the cases the approach works fine, it fails to identify the sentiment words whose polarity may be dependent on the context of usage. A classic example in sentiment analysis is the word *unpredictable* which has a clearly negative polarity if used in car domain: *The steering wheel of the car is unpredictable.* But is found to have a positive polarity in the movie domain: *The ending of the movie is unpredictable.* As a result, it can be stated that the context of using the word plays a major role in determining its polarity. The method followed in the paper is distinct in a way that it does not just follow the polarity of the words as given in the sentiment lexicon; instead, it computes it from the occurrence frequency of the sentiment word in the review dataset. The steps followed are as follows:

All adjectives are extracted from the POS tagged data. It need not be the case that all adjective words carry some sentiment meaning. For example, if we have the phrases:

front flash and *good flash*

in our reviews, POS tagging will tag both as:

front_JJ flash_NN and *good_JJ flash_NN*

As can be seen, both examples get reduced to Adjective(JJ)-Noun(NN) phrases. The phrase *good flash* carries some sentiment (positive) associated with it, whereas *front flash* does not. Hence, to identify whether an adjective is a sentiment word or not, we first make use of a sentiment lexicon [9] to find the presence of a specific adjective in the sentiment lexicon. If the adjective is present in the sentiment lexicon, we consider that word to be a sentiment word and vice versa. The sentiment lexicon is used only to identify if an adjective is a sentiment word or not. It is not used to actually identify the polarity of the sentiments.

The review dataset has a numeric rating associated with every review given by the reviewers in addition to the review text. The rating is in the form of stars and it ranges from 1 star to 5 stars. We group the reviews according to the number of stars associated with the review. To find the polarity of the sentiment words, we take the sentiment words extracted using the sentiment lexicon and find their occurrence across all five groups of reviews. If any sentiment word occurs the majority of the times in 4/5 star reviews, we consider that word to be extremely positive. If any sentiment word occurs the majority of the times in 1/2 star reviews, we consider that word to be extremely negative. If any word occurs an equal number of times in both, positive and negative reviews, we consider that word to be neutral in polarity. Hence, using this method we get the context-based sentiment polarity of the words.

4 Implementation and Experimental Results

The proposed work is implemented using Python programming language [12]. The preprocessing operations and POS tagging is carried out using the Natural Language Processing Toolkit (NLTK) [3]. Gensim topic modelling library [11] is used for LDA implementation.

The dataset for the study presented is sourced from Amazon [8]. It is a collection of Amazon reviews on the topic *Cell Phones and its related Accessories*. The first part of the work consists of the use of LDA to extract groups of words forming similar topics so that whenever the user starts Tweeting about a topic, all the related words get displayed to him/her to help him make his Tweet more informative. The optimal number of topics for the stated dataset was found to be 5 with five words in each topic. The algorithm was run for 10 passes. As we increased the number of topics and words, it is observed that the topic clusters lost their coherence; hence, we restrict the number of topics and number of words to 5.

The topics along with the words obtained are displayed in Table 1. Topic 1 corresponds to battery or charging. Topic 2 describes the case or the protection of the phone. The third topic can be estimated to vaguely describe the protector of the screen. The fourth topic is about cables or ports of mobile phones. The fifth topic, as can be seen, describes the overall quality of the phone. Hence, whenever a user types any one word from the topic, we can deliver him suggestions with other words in the same topic to help him make his tweet more informative.

The next part of the work deals with identifying sentiment words with the correct sentiment orientations in the concerned domain. Table 2 displays the normalized occurrence frequency of sentiment words across various review categories. As can be seen words like *poor, less, cheap* have higher occurrence frequency in 1/2 star

Table 1 Topics discovered using LDA

Topic 1	Battery	Device	Phone	Charger	Charge
Topic 2	Case	Protect	Color	Plastic	Phone
Topic 3	Protector	Screen	Thing	One	Part
Topic 4	Cable	Port	Car	Cord	Tip
Topic 5	Quality	Product	Time	Price	Sound

Table 2 Normalized occurrence frequency of sentiment words across review categories

1 Star	2 Star	3 Star	4 Star	5 Star
Less 1	Right 1	Bad 0.57269	Last 0.622323	Excellent 1
Poor 1	Second 0.452949	Cheap 0.355925	Clear 0.592842	Happy 1
Second 0.547051	Bad 0.297234	Clear 0.277108	Good 0.422779	Perfect 0.800259
Bad 0.42731	Cheap 0.267953	Full 0.235836	Little 0.301183	Good 0.703959
Cheap 0.376122	Last 0.170843	Good 0.172343	Nice 0.281524	Great 0.695143
Last 0.206834	Clear 0.13005	Hard 0.15393	Hard 0.27927	Easy 0.648894
First 0.095487	Better 0.079944	Better 0.153892	Long 0.266587	Best 0.640565
Better 0.064655	First 0.077419	First 0.114706	Easy 0.260472	New 0.637654
New 0.0644	Hard 0.077419	Little 0.107598	Best 0.248526	Long 0.632026
Good 0.053394	Good 0.070304	Nice 0.107075	Better 0.248526	Full 0.614961
Full 0.052943	New 0.053343	Long 0.101387	Full 0.235836	Nice 0.55073
Hard 0.052006	Nice 0.040275	Easy 0.090634	First 0.228803	Little 0.52144
Little 0.030312	Little 0.039468	New 0.085095	New 0.203437	First 0.508044
Nice 0.020395	Easy 0.029232	Best 0.069352	Perfect 0.199741	Better 0.452983
Great 0.018186	Great 0.027475	Great 0.060453	Great 0.198784	Hard 0.437375

reviews, hence, it can be inferred that these words are negative in polarity. On the other hand, words like *excellent, happy, great* have higher occurrence frequency in 4/5 star reviews; conforming their positive polarity.

5 Conclusion and Future Work

The work presented a method to assist users in the task of Tweeting informative messages in the required character length to make Twitter easier for the general public. The LDA algorithm used for topic modelling groups related words belonging to a common topic together from the dataset which can be displayed to the Twitter users as help while he/she is typing the Tweet. The problem of different words having different sentiment orientations in different domains is also taken care of since the method doesn't totally depend on sentiment lexicon to find sentiment orientations. Instead, the method computes the sentiment orientations of words based on the occurrence frequency in various classes of reviews. The future work consists of the development of the entire system and a live user trial to find the effectiveness of the system in the real world.

Acknowledgements This publication is an outcome of the research work supported by Visvesvaraya PhD Scheme, MeitY, Govt. of India (MEITY-PHD-2002).

References

1. Alhabash, S., Ma, M.: A tale of four platforms: Motivations and uses of facebook, twitter, instagram, and snapchat among college students? SAGE Publications (2017)
2. Amazon: https://www.amazon.com/
3. Bird, S., Loper, E.: Nltk: the natural language toolkit. In: Proceedings of the ACL 2004 on Interactive Poster and Demonstration Sessions (2004)
4. Blei, D.M., Ng, A.Y., Jordan, M.I.: Latent dirichlet allocation. J. Mach. Learn. Res. **3**, 1022–1993 (2003)
5. Curran, K., O'Hara, K., O'Brien, S.: The role of twitter in the world of business. Int. J. Bus. Data Commun. Netw. (2011)
6. Dong, R., Schaal, M., Ad Kevin McCarthy, M.P.O., Smyth, B.: Unsupervised topic extraction for the reviewer's assistant. In: International Conference on Innovative Techniques and Applications of Artificial Intelligence (2012)
7. Dong, R., Schaal, M., O'Mahony, M.P., Smyth, B.: Topic extraction from online reviews for classification and recommendation. In: Twenty-Third International Joint Conference on Artificial Intelligence (2013)
8. He, R., McAuley, J.: Ups and downs: modeling the visual evolution of fashion trends with oneclass collaborative filtering. WWW (2016)
9. Hu, M., Liu, B.: Mining and summarizing customer reviews. In: Proceedings of the ACM SIGKDD International Conference on Knowledge Discovery and Data Mining (2004)
10. ki Leung, C.W., Fai Chan, S.C., Lai Chung, F., Ngai, G.: A probabilistic rating inference framework for mining user preferences from reviews. World Wide Web **14**(2) (2011)
11. Řehřek, R., Sojka, P.: Gensim—statistical semantics in python. Statistical semantics; gensim; Python; LDA; SVD (2011)
12. Rossum, G.: Python reference manual (1995)
13. Taylor, A., Marcus, M., Santorini, B.: The penn treebank: An overview. In: Abeille A. (eds) Treebanks. Text, Speech and Language Technology, vol. 20. Springer, Dordrecht (2003)
14. Toutanova, K., Klein, D., Manning, C., Singer, Y.: Feature-rich part-of-speech tagging with a cyclic dependency network. In: Proceedings of HLTNAACL (2003)
15. Twitter: https://www.twitter.com/
16. Zhang, L., Liu, B.: Identifying noun product features that imply opinions. In: Proceedings of the 49th Annual Meeting of the Association for Computational Linguistics: Human Language Technologies (2011)

News Background Linking Using Document Similarity Techniques

Omkar Ajnadkar, Aman Jaiswal, P. Gourav Sharma, Chandra Shekhar, and Arun Kumar Soren

Abstract News readers still look forward to well-established news sources for a more accurate and detailed analysis of the news or articles. In this paper, we present a model that can retrieve other news articles that provide essential context and background information related to the news article that is being looked into. The abstract topics have been recognized using Latent Dirichlet Allocation. A similarity matrix has been created after the extraction of topics using two methods, cosine similarity and Jensen–Shannon divergence. This paper describes the implementation and gives a brief description of both the similarity methods. It also gives a comparative study of both methods.

Keywords Latent Dirichlet allocation · Cosine similarity · Jensen–Shannon distance · Document similarity

---

Please note that the LNCS Editorial assumes that all authors have used the western naming convention, with given names preceding surnames. This determines the structure of the names in the running heads and the author index.

O. Ajnadkar (✉) · A. Jaiswal · P. Gourav Sharma · C. Shekhar · A. K. Soren
Indian Institute of Information Technology, Kalyani 741235, West Bengal, India
e-mail: omkar@iiitkalyani.ac.in
URL: http://www.iiitkalyani.ac.in

A. Jaiswal
e-mail: aman137@iiitkalyani.ac.in

P. Gourav Sharma
e-mail: gourav@iiitkalyani.ac.in

C. Shekhar
e-mail: pavuluri_bt16@iiitkalyani.ac.in

A. K. Soren
e-mail: arun@iiitkalyani.ac.in

J. K. Mandal et al. (eds.), *Computational Intelligence and Machine Learning*, Advances in Intelligent Systems and Computing 1276,
https://doi.org/10.1007/978-981-15-8610-1_10

1 Introduction

In today's time, we have to keep ourselves notified with all the incidents happening around us. Keeping ourselves informed with the latest news around us shapes the way our mind thinks. To analyze an incident properly and to get to a conclusion and consequences, the reader must know the complete details of the incident. People who find the news interesting also look forward to a similar type of blogs and articles which report incidents that have occurred in the past. Despite the wide range of internet resources, the news seekers still look forward to a well-established system that would provide a more accurate and detailed analysis of the news events. Given a news article, the model should be able to find past related news articles or blogs related to the present news article. This background linking of information help users to find similar events that have occurred in the past, which in turn would help the reader to analyze the events in a better way. Much research is in progress in the field of information retrieval and natural language processing to develop a system of this kind. We have worked to develop a similar model. This model predicts similar articles related to the present news articles. The news articles are first tagged with topics that describe the news, i.e., sports, national, international, crime, etc. After assigning the news articles with the topics, we consider the topics of the news article of which the similar articles are to be found out. We find out the similarity of the topics with the topics of the other documents by creating a similarity matrix. This similarity matrix helps in measuring the similarity of topics for a given document with respect to others. Thus, we have the articles and blogs related to each other in a sorted order based on the score given by the similarity matrix.

2 Related Work

News background linking is the way to provide users context and background about the news they are reading now. Various rule-based approaches are previously used to find the text similarity and in turn to link related news articles.

In finding and linking incidents in news [1], authors use the clustering of text passages followed by pre-built rules for text linking. They also try to define the global score function to solve the optimization problem. In the paper Temporal Linking of News Stories [2], authors present a framework to link the past news articles of any article based on temporal expressions and other pieces of information such as persons mentioned and keywords. As many such methods work based on pre-built rules, they can become inefficient if the dataset size increases. Zhou Tong and Haiyi Zhang [3] did a text mining research and performed two experiments. In the first experiment, they applied LDA-based topic modeling on Wikipedia articles building a topic document model that provides a solution in searching, exploring, and recommending articles. The second experiment was conducted on twitter user's tweets which builds a model over Twitter user's interest. More recently, Minglai Shao

Fig. 1 Process flow

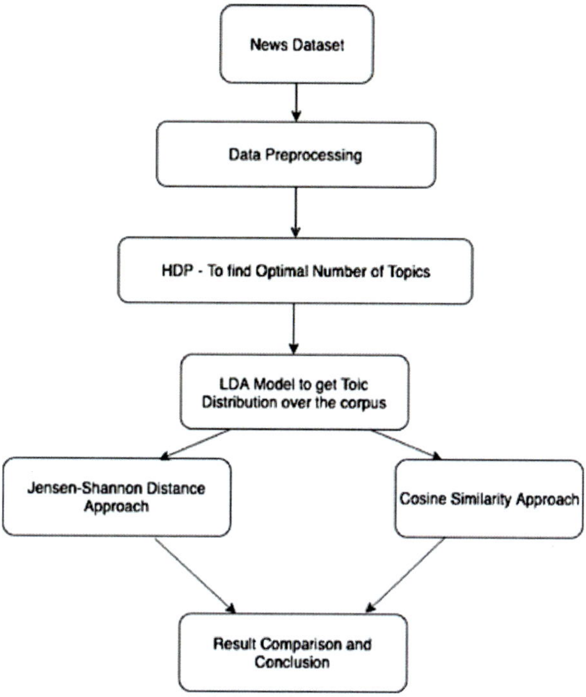

and Liangxi Qin [4] proposed that the LDA model + Jensen–Shannon (JSD) cannot differentiate semantic associations among text topics in computing text similarity. To this method, they introduced word co-occurrence analysis. Results show that the LDA model + JSD along with word co-occurrence analysis effectively improves text similarity computing result and text clustering accuracy (Fig. 1).

3 The Proposed Framework

3.1 Dataset

For the implementation of news background linking, we have used the TREC (Text Retrieval Conference) Washington Post Corpus. This dataset contains information regarding 608180 documents dated from January 2012 to August 2017. These articles are separated into six files that are present in JSON format. The dataset includes the details of all the news articles which include the title of the news article or blog post, the author's name, date of publication, section headers, article content, and links to embedded images and multimedia.

3.2 *Data Preprocessing*

As the data is obtained from the Washington Post Corpus, it is already spell checked and without grammatical errors. Still, as the dataset is in the format of the HTML dump, we pre-process the data before applying any further model for similarity checking. We perform several steps in series on the dataset as follows:

- HTML decoding and removal of images to get only news content
- Removal of links and numbers from text
- Contraction Expansion and negation handling
- Punctuation removal and lowercase transformation
- Stemming and tokenization.

We performed this pre-processing on both the news title and new content to give clean data as input the LDA model used for topic modeling.

3.3 *Topic Modelling*

Topic Modeling [5] is an unsupervised method of finding the most relevant topics from a collection of documents. The main idea is to represent documents as vectors of features (distribution of topics and words) where the distance between these features is used to compare them. Various algorithms like LDA, LSI, and NMF are used for the topic modeling. We use the method of LDA for the topic modeling in this paper.

Hierarchical Dirichlet Process (HDP) For the training LDA model, we have to assume several topics across the entire corpus. As the corpus is large enough, it is inefficient to decide manually. We used the Hierarchical Dirichlet Process (HDP) [6] approach and learns the correct number of topics from the given data. The model was learning some number of topics, we figured it out by applying the HDP model on the dataset. For the dataset, as the number was 20, we proceeded with it for the LDA model (Fig. 2).

Latent Dirichlet Allocation (LDA) LDA [5] is a topic modeling technique in which each document corresponds to several topics and each topic corresponds to several words. What exactly LDA does is

- LDA helps us getting the different sorts of topics in our corpus.
- LDA gives a better understanding of the type of documents in our corpus, i.e., news, magazines, stories.
- LDA identifies the most important words in a corpus, document similarity, which is all we want.

Here [7],

- M is the total number of documents
- N is the total number of words per document

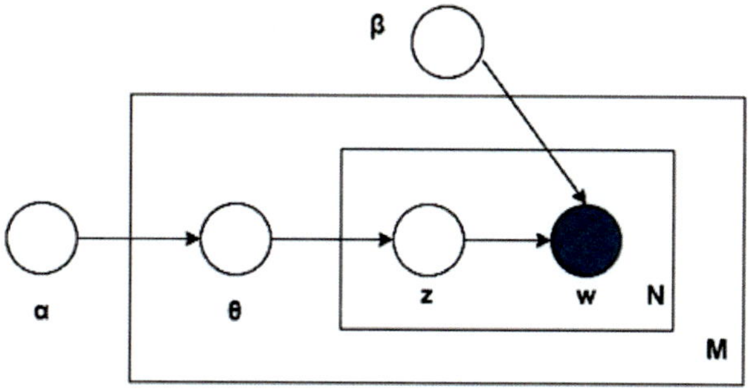

Fig. 2 Graphical model representation of LDA

- w are observed words in document i
- z gives the topic for word j in document i
- θ gives the topic distribution for a document i
- α and β is the document–topic density and topic–word density, respectively.

What exactly LDA gives:

- Topic distribution for each document. $[p(topic/document)]$
- Word distribution for each topic. $[p(word/topic)]$

We try to understand the model at the highest level. The model assumes that each document will hold several topics and the words in each document contribute to these topics.

3.4 Document Similarity

For predicting the news articles with the maximum similarity to the present article, we used two different techniques that used the trained LDA model as described in the above section.

Jensen–Shannon Distance Approach

After the training of the LDA model, on one side, each document in our corpus gives the topic distribution, and each topic gives the word distribution. On the other side, using the trained LDA model, we get the topic distribution of the test document, and our goal is to get the most similar documents in the corpus. Now, we get the topics vector and these vectors are used to measure the distance between the topics vector of a test document with the topics vector of all the documents in the corpus. This distance measured is known as the Jensen–Shannon distance metric. Jensen–Shannon distance is calculated by comparing the topic distribution of test document and the topic distribution of documents in the corpus. The formula of JSD [3] is obtained by the symmetrization of Kullback–Leibler Divergence(KLD). KLD's [4] had a limitation. KLD is the measure, which states the difference between a probability distribution and another reference distribution. It lacks symmetry and this is resolved by JSD.

For discrete distributions P and Q, the JSD [3] is defined as follows:

$$JSD(P \,||\, Q) = \frac{1}{2}D(P \,||\, M) + \frac{1}{2}D(Q \,||\, M)$$

where $M = \frac{1}{2}(P \,||\, Q)$
and D = Kullback–Leibler divergence

$$D(P \,||\, Q) = \sum_i P(i)log\left(\frac{P(i)}{Q(i)}\right)$$

$$JSD(P \,||\, Q) = \frac{1}{2}\sum_i \left[P(i)log\left(\frac{P(i)}{\frac{1}{2}(P(i)+Q(i))}\right) + Q(i)log\left(\frac{Q(i)}{\frac{1}{2}(P(i)+Q(i))}\right)\right]$$

The Jensen–Shannon Distance is given by

$$\sqrt{JSD(P \,||\, Q)}$$

We calculated the Jensen–Shannon distance for each document in the corpus and the test document. The value of Jensen–Shannon distance metric lies between 0 and 1. If the value is close to 0, then it says that two documents are more similar and if the value is close to 1, we can consider that the two documents are not at all related with each other. Finally, we set some threshold value between 0 and 1, say for example, 0.3 which signifies that group all the documents which are at least 70% similar to the test document. This threshold value indicates the degree of similar documents that we want.

Cosine Similarity Approach

The cosine similarity (CS) approach is used to find the degree of similarity between two vectors by measuring the cosine of the angle between the documents. By treating the topic distribution of each news as a vector, we use cosine similarity approach to find similarity between them.

CS between vectors A and B:

$$CS(A, B) = \cos(\theta) = \frac{A \cdot B}{\|A\| \; \|B\|} = \frac{\sum_{i=1}^{n} A_i B_i}{\sqrt{\sum_{i=1}^{n} A_i^2}\sqrt{\sum_{i=1}^{n} B_i^2}}$$

Mathematically, CS measures the cosine of the angle between two non-zero vectors, which is a representation of the orientation of one vector concerning others and not the magnitude. If the CS is 1, then the vectors are said to be closely related. The CS of 0 represents vectors orientated at 90 degrees, while CS of -1 is for the vectors that are diametrically opposite. As the angle between the vectors decreases, the CS value increases.

To find a document which is similar to the dataset for a given document, the following approach is used:

- The topic distribution across the complete dataset is obtained from the process of topic modeling using LDA.
- Using the trained LDA model, we find the top five topics for the given document. This is the document for which we need to find other similar documents.
- We iterate over other documents and find their topics using a trained LDA model. CS is used as an evaluation metric to find the similarity of these topics with respect to the given document.
- As the value of cosine similarity increases from −1 to 1, it suggests that the documents similarity increases. The document having maximum cosine similarity for the topics is the one which is most similar.

For example, the document with topics *obama*, *president* and *republicans* is more similar to the one with topics *senate*, *whitehouse* and *campaign*, rather than the one with *health*, *hospital* and *patients*.

4 Results and Analysis

As it was not possible to run the above model of finding similar news on the entire dataset due to its size and unavailability of ground truth, we selected a subset of the dataset with 400 articles, primarily focused on the domains of sports, politics and technology. We then manually labelled each news with three most similar news from the above subset by discussing with domain experts. For the comparison purpose, we assigned a score on range 0–3 to each news article, based on how many predicted news articles by the model matches with the labelled articles for it. While 0 means no predicted news is present in the manual labels, 3 is the score for the news whose all predictions match with the manual labels, We can analyze the output of both methods by the following example. Table 1 describes ground truth for the given news.

As the CS approach predicts two news articles out of three correctly in Table 2, we assign a linking score of 2 out of 3 to this news article.

As the Jensen–Shannon distance approach predicts all news articles correctly in Table 3, we assign a linking score of 3 out of 3 to this news article.

Table 1 Labeled similar news for the given news

Given news: bubbly planet venus starts off new year	
Labeled similar news for the given news	
1	Sky watch: vivacious venus returns
2	Jupiter bows out as venus takes center stage
3	For valentine's day, venus and jupiter snuggle up

Table 2 Cosine similarity approach

	Predicted news	Similarity score (%)
1	Sky watch: vivacious venus returns	54
2	Jupiter bows out as venus takes center stage	50
3	Williams sisters advance to gold medal match in doubles	39

Table 3 Jensen–Shannon distance approach

	Predicted news	Similarity score (%)
1	Sky watch: vivacious venus returns	58
2	For valentine's day, venus and jupiter snuggle up	53
3	Jupiter bows out as venus takes center stage	44

Table 4 Comparison between similarity approaches

Method	Average linking score
Cosine similarity approach	2.1
Jensen–Shannon distance approach	2.4

We repeat this process for all the 400 news articles using both methods and calculate the average linking score for them.

As we can observe in Table 4, the average linking score of the Jensen–Shannon distance approach is greater than the CS approach. So, the Jensen–Shannon approach works better than the CS approach for linking and finding similarity between long articles like entire news content.

5 Conclusion

In this paper, we discuss and analyze two different methods to get the related documents to the given document on the news articles dataset. We also discuss the critical ways in which these two methods differ. The document similarity task using topic modelling is enabling to get background contextual information about the given news, giving a detailed analysis of both the news and related topics. The automated background news linking using the image and video content along with the text data and author information are considerable problems for the research in this domain.

References

1. Feng, A., Allan, J.: Finding and linking incidents in news. In: Proceedings of the Sixteenth ACM Conference on Conference on Information and Knowledge Management, pp. 821–830. CIKM 2007. ACM. New York, NY, USA (2007).
2. Bögel, T., Gertz, M.: Time will tell: temporal linking of news stories. In: Proceedings of the 15th ACM/IEEE-CS Joint Conference on Digital Libraries, pp. 195–204. JCDL '15. ACM, New York, NY, USA (2015)
3. Tong, Z., Zhang, H.: A text mining research based on lda topic modelling. vol. 6, pp. 201–210 (2016)
4. Shao, M., Qin, L.: Text similarity computing based on lda topic model and word co-occurrence. In: 2014 2nd International Conference on Software Engineering, Knowledge Engineering and Information Engineering (SEKEIE 2014). Atlantis Press (2014)
5. Arun, R., Suresh, V., Veni Madhavan, C.E., Narasimha Murthy, M.N.: On finding the natural number of topics with latent dirichlet allocation: Some observations. In: Zaki, M.J., Yu, J.X., Ravindran, B., Pudi, V. (eds.) Advances in Knowledge Discovery and Data Mining, pp. 391–402. Springer, Berlin (2010)
6. Teh, Y., Jordan, M., Beal, M., Blei, D.: Hierarchical dirichlet processes. Mach. Learn. **12**, 1–30 (2006)
7. Blei, D., Ng, A., Jordan, M., Lafferty, J.: J. Mach. Learn. Res. **3**, 993–1022 (2003). submitted 2/02; published 1/03 latent dirichlet allocation. 02 2003

GRSS

Formation of Straight Line By Swarm Robots

Arijit Sil and Sruti Gan Chaudhuri

Abstract This paper proposes a distributed algorithm for computational mobile entities popularly known as swarm robots that operate and move in continuous space to arrange themselves in a straight line. The robots, following the standard OBLOT [1] model, are identical, anonymous, homogeneous. They do not agree on any global coordinate system, but the direction and orientation of the X-axis. Each robot maintains a local coordinate system. The robots do not communicate through message passing. The robots perform look–compute–move cycle in repetition in semi-synchronous scheduling where the robots do not store any data for future cycles. The algorithm presented assures collision-free movements of the robots.

Keywords Swarm robots · Line formation · Limited visibility

1 Introduction

A swarm of robots is a group of autonomous mobile computational agents performing a task collectively. Formation of different geometric patterns is one of the most fundamental tasks of mobile robots in order to execute a job cooperatively. The objective of this mobile robot systems is often to perform patrolling, sensing and exploring in a harsh environment such as disaster area, deep sea and space without any human intervention. In this paper we address the pattern formation problem using a deterministic algorithm that, being executed separately by every robot present in the universe U, allows the entire set to participate in formation of a straight line, within finite time, ensuring collision-free movements in a semi-synchronous environment.

A. Sil
Meghnad Saha Institute of Technology, Kolkata, India

S. G. Chaudhuri (✉)
Jadavpur University, Kolkata, India
e-mail: sruti1947@gmail.com

J. K. Mandal et al. (eds.), *Computational Intelligence and Machine Learning*, Advances in Intelligent Systems and Computing 1276,
https://doi.org/10.1007/978-981-15-8610-1_11

Earlier Works: All reported line formation algorithms [7] for mobile robots considers that the robots as points and they are able to sense all other robots. A point robot neither creates any visual obstruction nor acts as an obstacle in the path of other robots. Czyzowicz et al. [3] extended the traditional weak model of robots by replacing the point robots with unit disc robots (fat robots). Only some solutions on gathering problem has been reported for fat robots [1, 9, 10]. Under limited visibility gathering is solved for point robots [6] and fat robots [2]. Dutta et al. [4] proposed a circle formation algorithm for fat robots assuming common origin and axes for the robots. Here, the robots are assumed to be transparent in order to avoid visibility block. However, a robot acts as an physical obstacle if it falls in the path of other robots. Datta et al. [5] proposed another distributed algorithm for circle formation by a system of mobile asynchronous transparent fat robots with unlimited visibility. Vaidyanathan et al. [11] used robots bearing a constant number of lights in arbitrary pattern formation.

Framework: The system is composed of a set $R = \{r_1, \ldots, r_n\}$ of n computational entities or robots, that operate in a continuous spatial universe U in which they can move. The robots follow the de facto standard OBLOT model and manifest the following characteristics:

- Each robot is provided with its own local memory and is capable of performing local computations with (infinite precision) in real arithmetic.
- They are anonymous,i.e. they do not possess any distinct identities that can be used to separate it during the computation.
- The robots are autonomous, i.e. they operate without a central control or external supervision.
- They are homogeneous, i.e. they all have and execute the same protocol, or algorithm.
- A robot is represented as an opaque disc with a unit radius which works both as a physical and visual obstruction to other robots.
- The robots do not adhere to any global coordinate system. Each robot considers its position as the origin of its local coordinate system. They agree only on the direction (i.e. +ve and −ve X-axis) and orientation of X-axis. Although, it follows that the robots also agree on the orientation of Y-axis but not necessarily on the direction of +ve and −ve side of Y-axis. Finally, the robots also agree on the length of unit distance. Here, the radius of the robots is considered as one unit.
- A robot can observe/sense the part of the universe U that falls within its visibility range rad_v (same for all robots); the visibility range is assumed to be limited
- Each robot operates in Look–Compute–Move (LCM) cycles; in each cycle, it observes its surroundings, then it computes a destination point and finally moves towards it.

1. Look: The robot takes snapshot of visible universe U. This operation is instantaneous and the result is a snapshot indicating the positions of all other robots within its radius of visibility expressed the in local coordinate system.

2. Compute: The robot executes the algorithm (again, same for all robots), using the snapshot of the Look operation as input. The result of the computation is a destination point.
3. Move: The robot moves towards the computed destination.

- The robots execute this cycle following semi-synchronous scheduling where the activation of the robots is logically divided into global rounds; in each round, one or more robots are activated and obtain the same snapshot.
- The robots are oblivious. At the end of a cycle, all obtained information (observations, computations and move) are erased.
- The robots do not know about the total number of robots present in the system.
- The robots are silent. They have no means of direct communication of information to other robots.
- The robots form a logical graph G(V,E). Every robot is a vertex v $\in$ V in G. There exists an edge e, e $\in$ E between the robots r_i and r_j if and only if they can see each other. Initially, the graph G is assumed to be connected which implies that every robot can see at least one other robot.
- Initially, the robots are stationary and the mutual distance between two robots (the distance between the centres) is $\geq$ 3 units.

2 Overview of the Problem

Let $R = \{r_1, r_2, r_3.\dots.r_n\}$ is a set of n robots under the model described in the previous section. The location of a robot $r_i \in R$, is represented by the location of its centre which is always the origin of the local coordinate system Z_r. The robots in R have to move in such a way that after a finite amount of time they arrange themselves on a vertical (with respect to X-axis) straight line. Regardless of the speed, the robots are assumed to have rigid or unlimited mobility, i.e. all robots always reach their destinations when performing Move. In this model, the robots always move towards the destination point strictly along a straight line and never follows a curved path. As the robots are viewed as solid circular discs with extent, multiplicity is not permitted which means no two robots can occupy the same place at the same time. Collision is deemed undesirable, is avoided algorithmically by designing protocols for every valid movement.

Definition 1 Each robot $r_i \in R$, can see up to the perimeter of a fixed circular area around itself. This radius of this circular area is called the *visibility range* of r_i and is denoted by rad_v, where $rad_v > 4$ unit. *Visibility range* is of same length for every robot in R.

Definition 2 The circle centred at robot $r_i \in R$ and having radius rad_v is called the *visibility circle* of r_i denoted by VC(r_i). r_i cannot make any observation beyond the perimeter of VC(r_i).

Definition 3 The set of robots present within VC(r_i) of $r_i \in$ R are called *neighbours* of r_i.

Definition 4 A point is a *free point* if there exist no parts of another robot around a circular region of radius one unit around this point.

Definition 5 A path of a robot from source to destination point is called *free path*, if the rectangular area having length as the distance between source to destination and width as two units, does not contain any part of another robot.

When a robot $r_i \in$ R becomes active, it enters into the first phase of the active cycle, i.e. Look state. In Look state, r_i takes a snapshot of the robots that are present within VC(r_i) and plots those robots in its local coordinate system Z_r. This set of robots form neighbours of r_i with respect to Z_r, neighbours r_i can be divided into eight distinct and non-overlapping (not considering r_i) sets. Refer to Fig. 1.

- Set A consists of the robots partially or fully present in the area of unit distance around the positive Y-axis but it does not contain robot r_i itself.
- Set C consists of the robots partially or fully present in the area of unit distance around the positive X-axis.
- Set F represents the robots partially or fully present in the area of unit distance around the negative X-axis and set I contains the robots partially or fully present in the area of unit distance around the negative Y-axis.
- Set B and D account for the robots that fall within the first and second quadrants of the local coordinate system, respectively (leaving the robots in A, C and I).
- The robots which are in the third and fourth quadrants of the coordinate system Z_r make the sets E and G, respectively (leaving the robots on A, F and I).

Towards the centre of the coordinate system the sets A, C, F and I overlap. But, this area is only occupied by r_i as the model does not permit multiplicity and thus does not affect the overall distribution of neighbours of r_i in different sets.

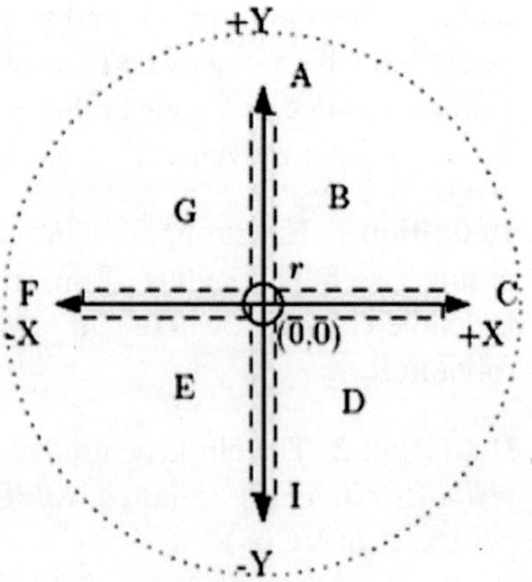

Fig. 1 Local view of a robot

Next, in the computation phase, r_i executes our algorithm to check whether it should make a move in this cycle or not. The algorithm considers all possible scenarios and determines whether r_i should travel or remain static and it also computes a destination point T for r_i. In the final phase called Move of the current cycle, r_i travels to the final destination point T. This movement of r_i to T adheres to the following criteria:

1. The visibility graph $G(V, E)$ remains connected.
2. T is a vacant point and the path to T from the initial location of r_i is a free path.

3 Description of the Algorithm

The algorithm, which is executed by all the robots separately under semi-synchronous environment, determines the length and direction of the movement of the concerned robot in the current cycle. The algorithm is divided into two subroutines. First subroutine, $NoMovement()$ identifies the cases where the robot shall not move, whereas the second subroutine $getDestination()$ computes the destination point T of the robot.

NoMovement

(r_i): r_i will not move under the following scenarios:

1. If r_i does not find any other robot in its visibility circle which happens if and only if there is only one robot.
2. If E ∪ F ∪ G is not empty.
3. If B ∪ C ∪ D is empty.

getDestination (r_i): This subroutine considers all the robots that are visible to r_i and ahead of it in the direction of positive X-axis (hereinafter referred as RIGHT). When B ∪ C ∪ D is not empty then the algorithm finds the nearest axis parallel to Y-axis that contains one or more robots from B ∪ C ∪ D. Let, ψ_{right} be that axis. The subroutine then considers four different scenarios.

Suppose, the robots partially or fully present in the area of unit distance around the axis ψ_{right} that are within the visibility range of r_i forms the set $R_{\psi right}$ (Fig. 2). The coordinate of r_i is taken as (0,0) w.r.t. its local coordinate system. We categorize different scenarios depending on the number and distribution of robots on ψ_{right}.

Scenario 1: If $R_{\psi right} \cap$ C is empty, (Fig. 3), which means none of the robots in $R_{\psi right}$ resides on X-axis, then the coordinates of the destination point T(x_T, y_T) is given by:
x_T = d + 0 [where d = distance along X-axis from current position of r_i to the intersection point between X-axis and ψ_{right}].
$y_T = 0$.

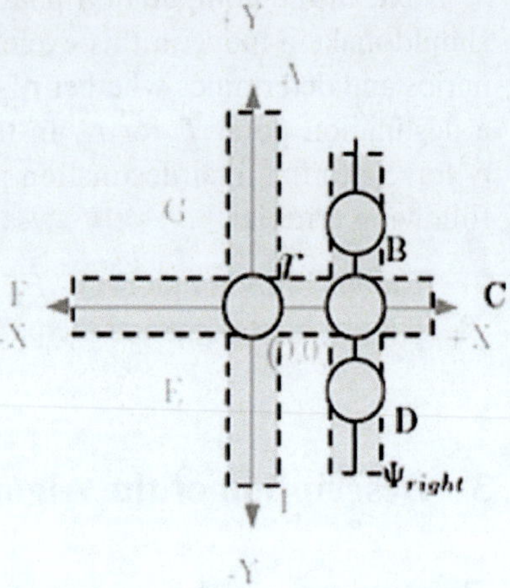

Fig. 2 An example of set $R_{\psi right}$

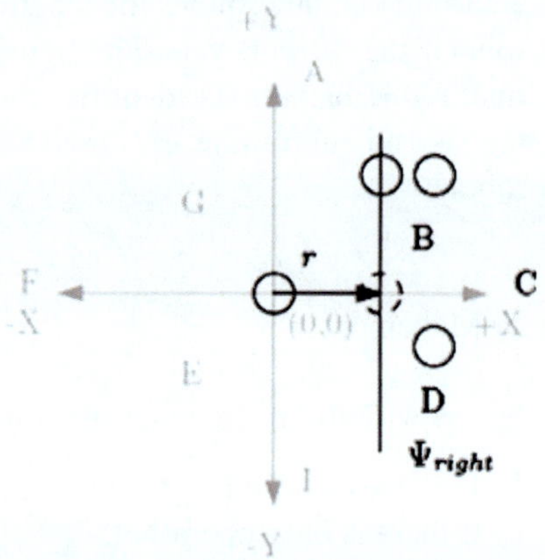

Fig. 3 Scenario 1

Scenario 2: If $R_{\psi right} \cap$ C is not empty and $R_{\psi right} \cap$ B is not empty and $R_{\psi right}$ $\cap$ D is empty (Fig. 4) then $R_{\psi right}$ has a robot on X-axis, it has got one or more robot that belongs to set B, but does not have any robot that belongs to set D. So, if r_i only moves horizontally, it will collide with the robot already present on the intersection point of ψ_{right} axis and X-axis as it is not a vacant point. So, its horizontal shift must be accompanied by a vertical shift in the negative Y direction. So, the coordinates of the destination point T is given by: $x_T = 0 + \text{d}$ [where d = distance along X-axis from current position of r_i to the intersection point between X-axis and ψ_{right}]. $y_T = 0 - 2$ X (1 unit distance) [where 1 unit distance = the radius of robots]. As the set D does not contain any robot on ψ_{right} axis, r_i does not have to face collision with any other robot as the path it follows along the hypotenuse of righ- angled triangle, with two other sides having length d and 2 X (1 unit distance), inside the second quadrant is a vacant path.

Scenario 3: If $R_{\psi right} \cap$ C is not empty and $R_{\psi right} \cap$ B is empty and $R_{\psi right} \cap$ D is not empty (Fig. 5), then $R_{\psi right}$ has a robot on X-axis, it has no robot that belongs to set B, but does have one or more robot that belongs to set D. So, if r_i only moves

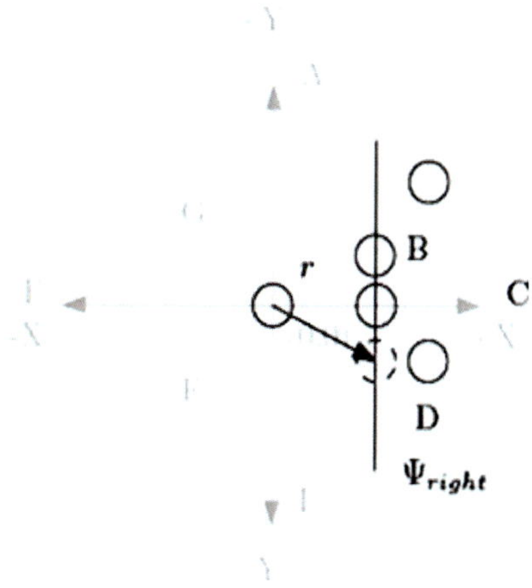

Fig. 4 Scenario 2

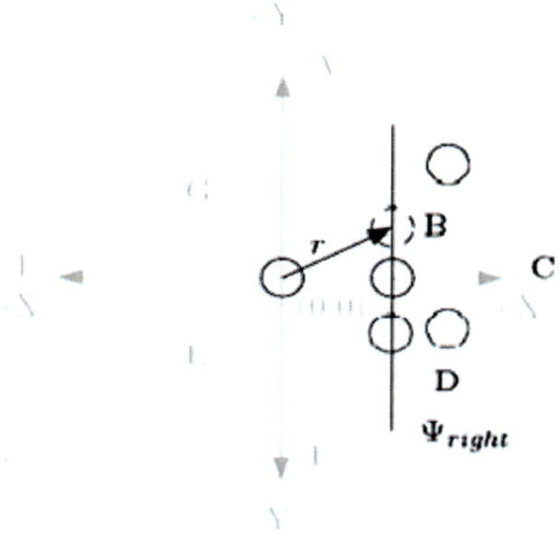

Fig. 5 Scenario 3

horizontally, just as in the previous scenario, it will collide with the robot already present on the intersection point of ψ_{right} axis and X-axis. So, its horizontal shift must be accompanied by a vertical shift towards the positive Y direction. So, the coordinates of the destination point T is given by:

$x_T = 0 + \text{d}$ [where d = distance along X-axis from current position of r_i to the intersection point between X-axis and ψ_{right}].

$y_T = 0 + 2$ X (1 unit distance) [where, 1 unit distance = the radius of robots].

As the set B does not contain any robot on ψ_{right} axis, r_i does not have to face collision with any other robot as the path it follows along the hypotenuse of right-angled triangle, with two other sides having a length and 2 X (1 unit distance), inside the first quadrant is a vacant path.

Scenario 4: If $R_{\psi right} \cap$ B, $R_{\psi right} \cap$ C and $R_{\psi right} \cap$ D none of these sets are empty (Fig. 6) then the robot r_i computes its destination point T as follows:

1. If A is completely empty and I is not empty, then r_i moves towards positive Y-axis. So the coordinates of the destination point T is given by:
   $x_T = 0.\ y_T = 0 + rad_v$

Fig. 6 Scenario 4

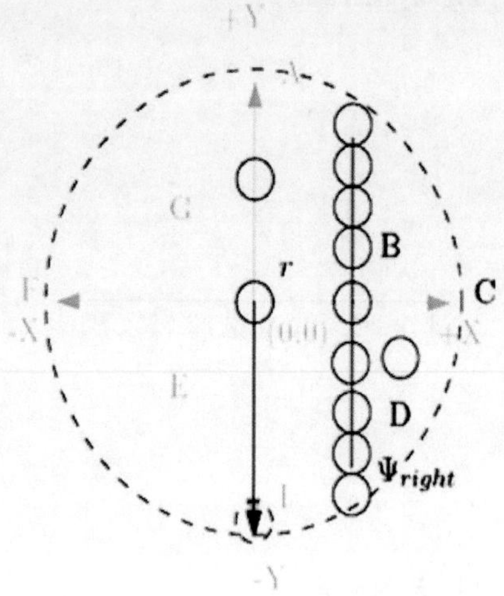

2. if I is empty and A is not empty, then r_i moves along negative Y-axis. So, the coordinates of the destination point T is given by:
   $x_T = 0.\ y_T = 0 - rad_v$.
3. If both A and I are empty, then r_i can move along either positive or negative Y-axis.
4. If both A and I are non-empty, then r_i does not move.

In this scenario, r_i only makes a move along positive or negative Y-axis when A or/and I is/are empty which makes the travel route a vacant path and as A or/and I is/are empty the destination point which lies on Y-axis and also is inside A or I is a vacant point. Therefore, the travel path of r_i is collision-free path.

4 Correctness

The robots successfully form a straight line in finite time using our proposed algorithm. The algorithm ensures the following features:

- The robots will move towards positive X axis with a goal to be placed on the global rightmost vertical axis RMA that contain any robot, on which the final straight line will be formed. This axis is static the algorithm assures that the robots will not cross RMA.
- The visibility graph G does not become disconnected. Otherwise, the robots in a particular component will have no knowledge about the robots in another component and a single straight line may not form.
- The robots do not collide.

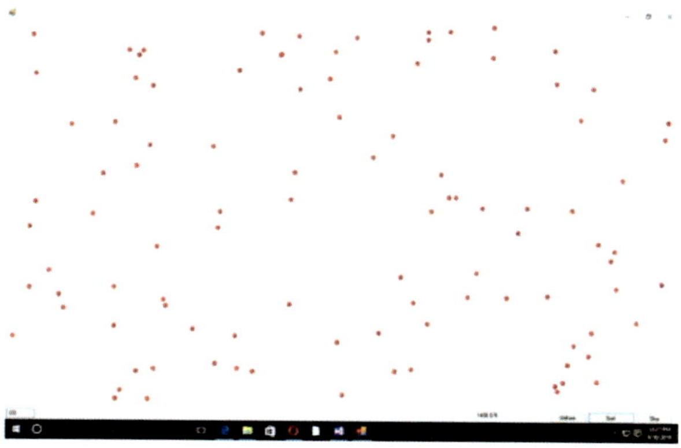

Fig. 7 Initial configuration

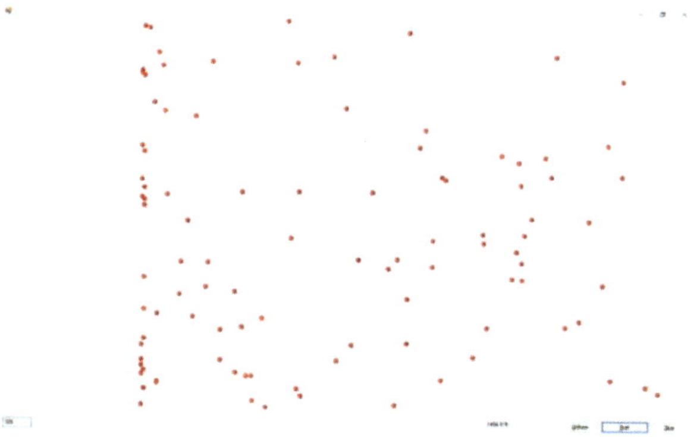

Fig. 8 Intermediate configuration

- The robots do not fall into deadlock to form the straight line as the intermediate configuration is such that at least one robot will be there to perform a movement towards RMA. This assures the finiteness of the algorithm.
- The semi-synchronous scheduling assures that the robots will not capture any moving robots. This eliminates the possibility of having obsolete positional data.

A simulation using Visual C# is successfully done to check the correctness of the algorithm. Figures 7, 8 and 9 show the snapshots of the simulation results of the positions of robots at initial, intermediate and final configurations, respectively.

Now we give the proofs of the above claims with the help of following lemmas.

Lemma 1 *The connectivity graph G remains connected.*

Proof r_i is in Scenario 1, robots in B or C or D will not move due to the presence of r_i according to our NoMovement( ) subroutine. Now, we will show that when r_i is moving to its destination it gets closer to the robots in B, C and D. Without loss of generality, let us prove this by taking the existence of any robot in B. The same arguments hold for the presence of robots in C or D.

Let s be the starting location of r_i (Fig. 10). Let there exists a robot r_b in B at point b. Let t be the destination of r_i. r_i moves along the edge *st* of the triangle δ *stb*.

Fig. 9 Final configuration

Fig. 10 An example of scenario 1 and connection preservation

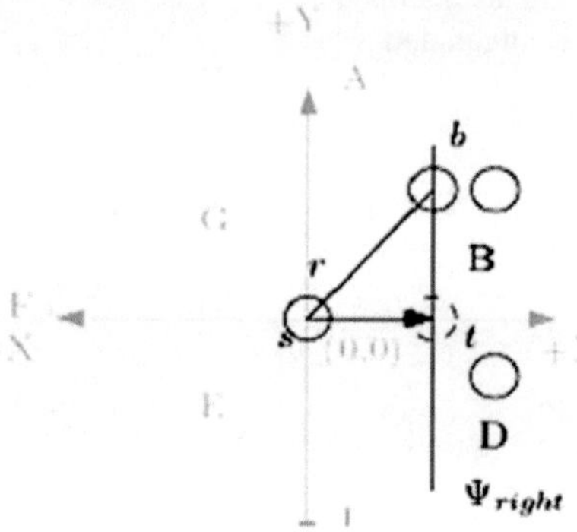

In the right-angled triangle *stb*, *tb* < sb (since *sb* is the hypotenuse). Hence, when r_i moves to t, it becomes closer to the robot present in B.

Using the similar argument we can prove that r_i is in Scenario 2 or 3 due to the movement of r_i, it becomes closer to the other robots present in B, C and D, Hence, there runs no chance to get disconnected with any robot within its visibility circle.

Now when r_i is in Scenario 4, no robot in B or C or D moves following NoMovement( ) subroutine. First, consider the case when I is empty. Then r_i moves along negative Y-axis by rad_v distance. Note that r_i in its new position is connected with the robots in D. Also note, the robots in D are connected with B, C and A. Using the same argument we can conclude that when r_i moves along positive Y-axis, as A is empty, for a distance of rad_v to its new position and remains connected with robots in B, C and D. Hence, G remains connected. □

Lemma 2 *The robots never collide*

Proof **(i) In Move phase, a robot r_i does not collide with any of the robots present within its visibility circle**:

In scenarios 1, 2 and 3, r_i moves to its nearest populated vertical line ψ_{right} only when there is a vacant point on it.

Consider scenario 1, r_i moves to the intersection point of +X-axis and ψ_{right}. This point is vacant according to the algorithm. The path towards this point from current position of r_i is also a free path as there is no vertical line between Y-axis and ψ_{right} containing any robot.

Consider scenario 2, r_i moves to the vacant point on ψ_{right} at B. The path towards this point from current position of r_i is also a free path as there is no vertical line between Y-axis and ψ_{right} containing any robot.

Scenario 3, is the mirror image of Scenario 2 where r_i moves to the vacant point on ψ_{right} at D through a free path.

Consider scenario 4, r_i moves up/down along Y-axis only if it can have a free path. Otherwise, it does not move and thereby nullifies the possibility of any collision.

(ii) In Move phase, a robot r_i does not collide with any of the robots present outside its visibility circle:

In scenarios 1, 2 and 3, r_i moves to a vacant point t on its nearest populated vertical line ψ_{right}. Now a robot $r_{outside}$ which lies outside the visibility circle of r_i cannot possibly compute point t for its destination point, as in order to do so $r_{outside}$ cannot be separated from r_i by more than 4 unit distance. But, as the fixed visibility circle radius for all the robots is at least $rad_v > 4 - unitdistance$, $r_{outside}$ will be having a destination point other than t. Consequently, r_i and $r_{outside}$ never collide.

In scenario 4, r_i moves only along the (+ve) or $(-ve)$ Y-axis only by a distance of 2 unit. So, it is not possible to collide with a robot $r_{outside}$, as it is at least 5-unit distance away from r_i. □

Lemma 3 *There exists always a robot which will move unless the robots in R has formed a straight line on RMA.*

Proof If a robot sees any robot at its RIGHT side, it will move. If it does not see any robot in B ∪ C ∪ D, it does not move. This is possible in the following two cases. (i)There is only one single robot. (ii)The robots have formed a straight line. The robots do not form a straight line, but there exists another robot in A or I which has the connectivity with the right side or left side of the Y-axis. For both the cases, there exists robot other than r_i, which will move. Hence, the lemma is true. □

Lemma 4 *If the robots do not form a vertical straight line yet, there is always a robot that will leave its Y-axis after a finite time and move in the +X direction.*

Proof The proof follows from lemma 3. Given a set of robots R on the 2D plane in its initial configuration, through our algorithm the robots are finally placed on the RMA and form the required straight line. □

Lemma 5 *Each robot will move closer to RMA in finite time interval.*

Proof If a robot finds any robot at its right side, it moves in +X direction following scenarios 1, 2 and 3 on to its ψ_{right}. As a result, robots do not stay idle for infinite time and is also guaranteed to reach to its ψ_{right} in finite time, i.e. closer to RMA. In scenario 4, r_i moves up and down along Y-axis if there exists a free path. When

r_i moves up/down its visibility circle also moves up/down and it covers a new set of robots. Eventually, r_i moves towards right and placed on the next vertical line nearer to RMA. If r_i does not move down, there exists another robot to move and, eventually, r_i gets its chance to move unless the straight line is already formed. □

Lemma 6 *None of the robots ever crosses the RMA of the set of robot.*

Proof Suppose there is a robot r_i which has crossed RMA. r_i can do that in two different ways. If r_i was initially on RMA then, it has left that axis to move to RIGHT or r_i was initially LEFT of RMA and has crossed it while going towards RIGHT. In the first case to leave RMA, r_i has to observe an axis containing robots towards RIGHT. But as r_i was sitting on RMA no such axis can exist and therefore it contradicts our assumption and therefore once on RMA, r_i cannot leave it anymore. In the second case, r_i was LEFT of RMA and in order to go past RMA, it has to find RMA to be the nearest axis containing robots towards RIGHT as there is no other axis with robots RIGHT of RMA. But if RMA is the nearest axis then the maximum horizontal shift would take r_i up to RMA and not beyond that and once it reaches RMA, r_i cannot leave it anymore. Therefore, the scenario contradicts our assumption. So, by contradiction, we can say that none of the robots ever crosses RMA. □

Lemma 7 *All the robots in R will be on the RMA in finite time.*

Proof According to Lemma 5, each robot r_i reaches its ψ_{right} in finite time. This implies that after a finite time there will be a configuration where RMA will be the ψ_{right} for every robot. After this configuration in finite time, all robots will move on RMA. □

Finally we can summarize the result in the following theorem:

Theorem 1 *A set of autonomous, oblivious, homogeneous, non-communicative, fat and opaque robots having limited visibility while operating under semi synchronous environment with one axis agreement, can form a straight line in finite time using collision-free movements.*

5 Conclusion

In this paper, we have shown that, for a set of weak and fat robots, there exists a solution to straight line formation problem under limited visibility in a semi- synchronous environment without collision. In future, we would like to address the same problem excluding axis agreement. Inspecting the solvability of identical issues in a fully asynchronous environment also poses an interesting challenge.

References

1. Agathangelou, C., Georgiou, C., Mavronicolas, M.: A distributed algorithm for gath-ering many fat mobile robots in the plane. In: Proceedings of the 2013 ACM Symposium on Principles of Distributed Computing, pp. 250–259 (2013)
2. Bolla, K., Kovacs, T., Fazekas, G.: Gathering of fat robots with limited visibility and without global navigation. In: Swarm and Evolutionary Computation. LNCS, vol. 7269, pp. 30–38 (2012)
3. Chaudhuri, S.G., Mukhopadhyaya, K.: Distributed Algorithms for Swarm Robots, Handbook of Research on Design, Control, and Modeling of Swarm Robotic, IGI Global, Copyright@2016, p. 26
4. Chaudhuri, S.G., Mukhopadhyaya, K.: Leader election and gathering for asyn-chronous fat robots without common chirality. J. Discret. Algorithms **33**, 171–192 (2015)
5. Czyzowicz, J., Gasieniec, L., Pelc, A.: Gathering few fat mobile robots in the plane. Theor. Comput. Sci. **410**(67), 481–499 (2009). Principles of Distributed Systems
6. Datta, S., Dutta, A., Chaudhuri, S.G., Mukhopadhyaya, K.: Circle formation by asynchronous transparent fat robots. In: International Conference on Distributed Computing and Internet Technology, pp. 195–207. Springer, Berlin (2013)
7. Dutta, A., Chaudhuri, S.G., Datta, S., Mukhopadhyaya, K.: Circle formation by asynchronous fat robots with limited visibility. In: International Conference on Distributed Computing and Internet Technology, pp. 83–93. Springer, Berlin (2012)
8. Flocchini, P., Prencipe, G., Santoro, N.: Distributed computing by oblivious mobile robots. Synth. Lect. Distrib. Comput. Theory **3**(2), 1–185 (2012)
9. Flocchini,P., Principe, G., Santoro, N.: Moving and computing models: robots, distributed computing by mobile entities. In: LNCS, vol. 11340, pp. 3–14 (2019)
10. Flocchinia, P., Prencipeb, G., Santoro, N., Widmayerd, P.: Gathering of asynchronous robots with limited visibility. Theor. Comput. Sci. **337**, 147–168 (2005)
11. Vaidyanathan, R., Sharma, G., Trahan, J.L.: On fast pattern formation by autonomous robots. SSS 203–220 (2018)

A Novel Technique to Utilize Geopolitical Risk as a Factor for Predicting Gold Price

Debanjan Banerjee and Arijit Ghosal

Abstract Accurate prediction of commodity prices by using geopolitical risk as a significant factor has been investigated by researchers and investors. The proposed work creates a model for the quantification of geopolitical risk by using country and event-specific economic impact as aspects. The work also utilizes panel data regression techniques on different gamuts of data to identify the features that have the most influence on the gold price. The work also contributes to developing the concept of event intensity which can be considered as a variable that helps us to investigate the potential impact on the economic sphere of the geopolitical event in a given time period.

Keywords Geopolitical risk · Linear regression · Machine learning · Event-intensity

1 Introduction

In the contemporary world, some political events such as incidents of war, terrorism do have a significant impact over economic events and they significantly impact commodity prices like gold or crude oil. However, the impact of all the events are not same, i.e., reinstatement of the US sanctions on Iran will not have the same impact on the worldwide commodity prices as India recalling her ambassador from Pakistan. The type of event is also by itself a very significant factor in influencing

D. Banerjee
Sarva Siksha Mission, Kolkata, India
e-mail: debanjanbanerjee2009@gmail.com

A. Ghosal (✉)
St. Thomas' College of Engineering and Technology, Kolkata, India
e-mail: ghosal.arijit@yahoo.com

J. K. Mandal et al. (eds.), *Computational Intelligence and Machine Learning*, Advances in Intelligent Systems and Computing 1276,
https://doi.org/10.1007/978-981-15-8610-1_12

the possible impact over many subsequent issues such as the economic output of a country. Until now most researchers have attempted to quantify the geopolitical risk in terms of total references to a given event. However, the gap in this particular approach to the quantification of geopolitical risk is that this approach does not take into consideration the potential economic impact of the particular geopolitical event as well as the fact that different types of geopolitical event may cause a variable impact on the economic front. Thus, there is the need to take into consideration the economic impact of the geopolitical events for quantification of the geopolitical events. The current work utilizes a novel method to quantify geopolitical risk while using constraints for geographical position and type of geopolitical event and thus forms the geopolitical risk model. The major difference between the current work and previous other works with respect to the quantification of geopolitical risk is that the current work takes into consideration the potential impact on gross domestic product (GDP) before and after a specific geopolitical event. All the data and features which are part of the work have been collected from the open source, i.e., various online sources like moneycontrol.com.

2 Related Work

The researcher community is concerned about the fact that whether and how much the political events of the day do impact the macroeconomics of a country. Baker et al. [2] built an economic uncertainty model and this proved that the US economic situation fluctuates regularly whenever important events like the US presidential elections or wars take place. Banerjee et al. [1] utilized the concept of geopolitical risk to predict the price of London Gold Price fix prices. Barro and Ursua [3], Barro [4], and Blum [6] have observed that unpredicted macroeconomic events do severely impact stock market prices and, correspondingly, they also impact the gold prices. Caldara and Iacoviello [7] first conceptualized and quantified the geopolitical risk factor and they did use the total amounts of references in news media for that purpose. The current work improves upon their previous work by introducing different aspects like country-specificness, event intensity, and potential economic impact of these geopolitical events to formularize geopolitical risk in an unique way. Das et al. [8] utilized panel data regression techniques to investigate the impact of FDI on labor productivity in the Indian banking sector. Jiaying and Stanislav [9] performed panel data regression on multivariate data to explore quantile properties of the same data. Saiz and Simhonson [5] has observed that specific newspaper items like terrorism and war do have significant impacts on the American economy.

3 Importance of Selecting Geopolitical Risk as a Feature

The present approach for quantifying geopolitical risk and analyzing its impact over gold price can be described in the following steps. This work considers geopolitical risk.

- Defining geopolitical risk: This is the very first step whereby this work defines geopolitical risk as a function of three important aspects. The country of origin of the event in question, the type of event, and most importantly the economic cost of the event. The economic cost could be considered in terms of possible fluctuation in the stock and commodity markets as well as in terms of domestic and foreign investments.
- Computation of geopolitical risk: The current work uses a formula whereby the geopolitical risk is computed in a daily manner involving the economic aspect of the impact of the geopolitical event. The location of the situation, the types of situation are taken into consideration while computing the geopolitical risk.
- The current work utilizes geopolitical risk as a feature and would like to observe the influence of the same with respect to the London Gold Price Fix.

Geopolitical Risk as a Feature Group The whole gamut of features that are having possibly significant influence over gold price can be categorized into four groups.

These groups are

- Supply-based features: These types of features are gold extraction indexes, the total production of gold production companies and countries, etc.
- Demand features: These types of features are GDP growth rate, BDI rates, various countries′ like the US, Russia, China central bank gold purchase figures, etc.
- Financial features: Different types of stock and commodity exchange NASDAQ and Dow Jones stock exchanges, currency exchange rates such as US Dollar–Chinese Yuan exchange rate, Euro–Yuan exchange rate, etc.
- Political features: Geopolitical risk can be categorized under this particular type of feature group. The potential challenges involve identifying and then quantifying factors that determine geopolitical risk.

Quantifying Geopolitical risk as a feature The current work computes the geopolitical index based upon the following assumptions:

- Geopolitical events in some countries cause more uncertainty than the rest of the countries and different types of geopolitical events have different types of uncertainty ramifications.

The current work utilizes the spirit of Cobb–Douglas formulations after context-based customizations to define geopolitical risk.

- The work assumes c_t as the total contribution of the GDP of a country to the world GDP in percentage at a given time t. This helps to categorize a given country based upon their overall importance to the global economy.

- The work assumes m_t as the total change of the country's GDP given a geopolitical event at a given time t. This is crucial to investigate the amount of impact the geopolitical event makes on a country's economic output over a period of time t.
- The work assumes v_t as the total change in the overall economic contribution of the country to the world economy at a given time t. This helps to quantify the economic impact of the geopolitical event to the contribution of the country to the global economy.

The equation can also be expressed as the following:

$$G_t = c_t^{\theta} m_t^{1-\theta} \exp v_t, \tag{1}$$

where G_t is considered as geopolitical risk and c refers to the origin of the country, whereas m represents the total monetary impact of the same event and v refers to the event and t represents time. Dividing both side of the Eq. 1 with m_t the following Eq. 2 can be obtained.

$$G_t/m_t = (c_t/m_t)^{\theta} \exp v_t. \tag{2}$$

In this Eq. 2, the ratio c_t/m_t will be referred further as $event\text{—}intensity$. So, we can formulate geopolitical risk as on the following Eq. 3 with this ratio.

$$G_t/m_t = (c_t/m_t, v_t). \tag{3}$$

The empirically estimable Eq. 4 thus can be derived by taking natural logarithm on both sides of Eq. 3 for a particular time t.

$$ln(G_t/m_t) = \exp(\ln(c_t/m_t)) + \exp(\ln(v_t)) + u_t, \tag{4}$$

whereas u_t represents the error term in the Eq. 4.

Once we complete the computation of the geopolitical risk as a feature by using the above equations, we then include the same alongside other features to investigate their potential influence over gold price by using regression procedures.

4 Experimental Results for the Regression Procedure

This work utilizes the panel data regression techniques with the help of the library *plm* from the R programming language libraries. The panel data regression techniques have been utilized since there are more than one variables in question and time is also an important factor in these equations. All the major experimental results involving all the major features on which the regression method was applied with have been depicted in Table 1.

The work performs regression based upon certain criterion.

Table 1 Comparison between geopolitical risk and other features

Feature	Slope	*p*-value	R2 error	F-statistic *p*-value
Geopolitical risk	1.683	4.585 e16	0.918	5.376 e–16
US GDP growth rate	3.683	0.6413	0.008	0.7615
China GDP growth rate	5.683	0.5198	0.009	0.5376
China central bank gold purchases	11.683	2.9785 e–16	0.818	3.459 e–16
NASDAQ stock exchange index	4.683	0.1467	0.678	0.1543
USD–CNY exchange rate	1.367	6.789 e–16	0.927	7.76 e–16

- The work uses a null hypothesis which assumes that there is no influence of the utilized feature over the gold price and the target variable in these regression procedures is London Gold Price Fix PM.
- The null hypothesis can be rejected based upon two following conditions when the prob. value and the F-statistic prob. value both are having a value less than 0.01 then only the null hypothesis is rejected and the alternate hypothesis, i.e., the utilized feature in question does have influence over the given target variable is accepted.

We observe by analyzing the data from regression statistics in Table 1 that the following features do influence the gold price as they all satisfy the same following condition:

- Geopolitical risk feature has Probability 4.585e–16 and F-statistic Probability 5.376e–16. Both of these values are less than 0.01.
- China central bank gold price purchases feature has Probability 2.9785e–16 and F-statistic Probability 3.459 e–16. Both of these values are less than 0.01.
- USD–CNY exchange rate feature has Probability 6.789e–16 and F-statistic Probability 7.76e–16. Both of these values are less than 0.01.

By utilizing these three features the following model for prediction of London Gold Price Fix can be constructed as depicted in the Eq. 5.

$$Y = 1.683 \times X1 + 11.683 \times X2 + 1.367 \times X3 + e \tag{5}$$

- Here, we refer London Gold Price Fix as Y, we refer Geopolitical risk as X1, Chinabank gold purchases as X2, USD–CNY as X3 and Random Error as e.

4.1 Comparison with Previous Work

The current work performs better when being compared with previous work that was performed by Banerjee et al. [2] which utilized the two-day lags of the three features

Table 2 Comparison between geopolitical risk and other features

Work	R^2 error
Banerjee et al. [2]	0.6046
Das et al. [8]	0.7046
Proposed approach	0.7358

of the model and Das et al. [8] which utilized the power of the panel data concepts. The details of the experiments in terms of the R^2 error have been depicted in the following Table 2. The newly constructed geopolitical risk index thus helps improve the accuracy of the work. So it can be observed from Table 2 that there is a significant improvement of performances after using the proposed approach.

5 Conclusion

This can be observed from the experimental results that using the aspects of country-specific events and their potential impacts of those events on the overall economic productivity of the country and to the overall economic contribution of the country to global productivity help to give a completely new insight into look into the question of geopolitical risk and this in result helps to improve the accuracy for the prediction of the gold prices.

References

1. Baker, S.R., Bloom, N., Davis, S.J.: Measuring economic policy uncertainity. Q. J. Econ. **131**(4), 1593 (2016)
2. Banerjee, D., Ghosal, A., Mukherjee, I.: Prediction of gold price movement using geopolitical risk as a factor. In: Emerging Technologies in Data Mining and Information Security, pp. 879–886. Springer, Singapore (2019)
3. Barro, R.J., Ursua, J.F.: Rare macroeconomic disasters. Ann. Rev. Econ. **4**(1), 83–109 (2012)
4. Barro, R.J.: Rare disasters and asset markets in the twentieth century. Q. J. Econ. **121**(3), 823–866 (2006)
5. Bloom, N.: The impact of uncertainity shocks. Econometrica **77**(3), 623–685 (2009)
6. Caldara, D., Iacoviello, M.: Measuring geopolitical risk. In: Working Paper, Board of. (2016)
7. Das, G., Ray Chaudhuri, B.: Impact of FDI on Labour Productivity of Indian IT Firms: horizontal spillover effects. J. Soc. Manag. Sci. **XLVII**(2) (2018)
8. Jiaying, G., Stanislav, V.: Panel data quantile regression with grouped fixed effects. J. Econ. **213**(1), 68–91 (2019)
9. Saiz, A., Simonsohn, U.: Proxying for unobservable variables with internet document frequency. J. Eur. Econ. Associ. **11**(1), 137–165 (2013)

Effect of Window Size on Fourier Space of CMAM30 Model Data Generated for Satellite Sampling

Subhajit Debnath and Uma Das

Abstract Canadian Middle Atmosphere Model (CMAM30) data is used to analyze the temperature variation from 1000 to 0.001 hPa from 2009 to 2010 to study the effect of window size on two-dimensional Fourier spectra. Mean temperature, stationary planetary wave (SPW1), and tidal components are extracted from the spectra using window sizes of ±2, ±10 and ±30 days and examined over the equator and mid-latitudes. The migrating diurnal tide, DW1, has maximum amplitude over the equator and increases with increasing altitude. SPW1, on the other hand, has maximum amplitude over mid-latitudes and is practically absent over the equator. Further, by using smaller window size short-term variabilities and finer variations of SPW1 and the tides are understood. The Fourier space of model data is independent of window size for satellite sampling and is thus concluded that large and non-overlapping windows are suited for generating a Fourier space for satellite sampling.

Keywords Atmospheric tides · Canadian middle atmosphere model · Fourier space · Satellite sampling

1 Introduction

Atmospheric tides are harmonics of solar days (24, 12, 8 h). Tides are generated by diabatic heating of the atmosphere and wave–wave interactions. Tides are classified into two types: (a) migrating tides which propagate westward with same phase speed as the Sun and (b) non-migrating tides which propagate westward or eastward or are stationary and have a phase speed that is different from the Sun [1–5]. Classical tidal theory [6, 7] suggests that the real atmosphere can be represented as a superposition of all these wave components over a background field. As the tides move

S. Debnath · U. Das (✉)
Indian Institute of Information Technology Kalyani, Kalyani 741235, West Bengal, India
e-mail: uma@iiitkalyani.ac.in

S. Debnath
e-mail: subhajitd@iiitkalyani.ac.in

J. K. Mandal et al. (eds.), *Computational Intelligence and Machine Learning*, Advances in Intelligent Systems and Computing 1276,
https://doi.org/10.1007/978-981-15-8610-1_13

vertically upward, tidal amplitudes increase with altitude due to decreasing density. In the northern polar region, the amplitude of diurnal tide DW1 (Diurnal Westward propagating tide of wave number 1) is maximum between April and October and also maximum in the southern polar region between October and April. The amplitude of DS0 (Diurnal Stationary tide) is maximum in both polar region during May–September [8].

Variability of tides in atmosphere has been established by investigations of model and satellite datasets. TIMED (Thermosphere, Ionosphere, Mesosphere Energetics, and Dynamics) is one of the satellites, whose data has provided many insights into tidal variability. Due to the yaw of the TIMED satellite, which is in a near Sun-synchronous orbit, it can cover all 24 h local times over a given location in approximately 60 days [9]. So, by analyzing this data, we only get the average picture of the tidal field. An important assumption here is that the background temperature is not varying during this 60-day period. However, it is not true in reality and so the background temperature can alias into the diurnal migrating tide [2]. Aliasing can also take place between Stationary Planetary Wave of wavenumber 1 (SPW1) and non-migrating tides DS0 and DW2 [10]. Other satellites like FORMOSAT-3/COSMIC (Formosa Satellite-3/Constellation Observing System for Meteorology, Ionosphere, and Climate) with a different sampling pattern are capable of addressing these aliasing problems and extracting short-term tidal variability in the Earth's atmosphere. COSMIC also has a limitation as it covers altitudes up to ~50 km only from the surface of the Earth.

To identify aliasing components ground truth is absolutely necessary, which is absent while working with real satellite data and hence model data is used. Assuming model data as ground truth, the same can be subsampled at locations and times of satellite observations to obtain a new satellite sampled model dataset. This new dataset may then be spectrally analyzed to obtain the different wave components and investigated to understand the underlying aliasing problems.

The current study focuses on the various details involved in the subsampling of the model data. Canadian Middle Atmosphere Model (CMAM30) model temperature data is spectrally analyzed using a two dimensional fast Fourier transform and the effect of window size is investigated in generating a suitable satellite sampled dataset. Fourier space subsampling is studied and compared with the regular linear interpolation method.

2 Data and Method

CMAM30 estimates the chemical and dynamical evolution of the atmosphere [11–13]. There are two types of datasets: (a) regular CMAM30 dataset (up to around 95 km) which is concentrated on the troposphere–stratosphere–mesosphere region. (b) Extended CMAM30 (up to around 200 km) dataset which is concentrated on the Mesosphere and lower thermosphere region [14–16]. In the current study, CMAM30 temperature data is used that is provided every 6 h from 1979 to 2010 from 1000 hPa to 0.001 hPa. The longitude by latitude grid size is 3.75 × 3.75. Temperature data, $f(t,\lambda)$, is analyzed for the period from 2009 to 2010 spectrally to extract diurnal tides, both migrating and non-migrating tidal components. A two-dimensional fast Fourier transform (Eq. 1) is applied in longitude (λ) and time (t) to get the Fourier spectrum, $F(f, s)$, as a function of frequency (f) and wave number (s), using different window sizes, of ±2, ±10 and ±30 days.

$$F(f, s) = \frac{1}{N} \sum_{t} \sum_{\lambda} f(t, \lambda) e^{-i2\pi(ft+s\lambda)} \quad (1)$$

where N is the total number of data points considered. In the spectra, positive wavenumbers indicate westward propagating waves and negative wavenumbers indicate eastward propagating waves [8]. Amplitudes of the diurnal tides DW1, DS0, DW2, and SPW1 are extracted and investigated over the equator and mid-latitudes to understand the effect of the size of window used in the 2DFFT. Further, the Fourier space is sampled for non-grid longitudes and times and inverted to real space to mimic satellite sampling. The difference between sampling in Fourier space is compared with linear interpolation in real space and the results are discussed in the ensuing sections.

3 Result and Discussion

Mean temperature and amplitudes and phases of SPW1, DS0, DW1, and DW2 from 2009 to 2010 over the equator are shown in Fig. 1. Mean temperature shows the stratopause at 1 hPa/50 km which varies semi-annually. The mesopause is not visible in the plot and is above 0.001 hPa/95 km. SPW1 is practically absent over the equator at all altitudes; the phase plot also does not indicate any significant SPW1 variation. The diurnal tides are more important here. DW1 peaks over the equator in the mesosphere, with amplitudes larger than 4 K. In the stratosphere, DW1 amplitudes of 1 K are observed. The vertical wavelength (as seen in the phase plot) is approximately 25 km, and consistent throughout the study period. This property is similar to what has been observed in earlier studies. DS0 and DW2, on the other hand, are present only in the mesosphere. The vertical wavelength of these non-migrating tides is similar to the migrating tide, i.e., 25 km [17–19]. Interestingly, DS0 and DW2 seem to be mutually exclusive in the upper mesosphere. DS0 peaks during late winter and spring equinox and DW2 peaks during autumn and early winter. With the reduction in window size to ±10 and ±5 days, the properties of the mean temperature and the waves gave the same results but with finer variations (figures are not shown here for equator).

Figure 2 shows the same details as in Fig. 1, but over mid-latitudes (~60°N). The annual variation of the mean temperature is very prominent with warm stratopause and cold mesopause during summer. SPW1 shows significant amplitudes up to 12 K

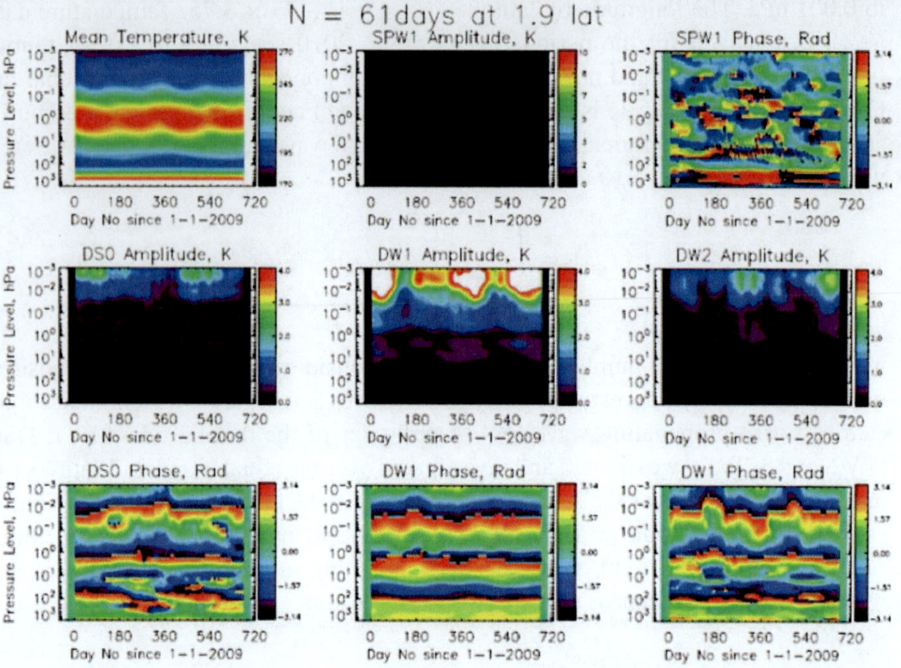

Fig. 1 Mean temperature, stationary planetary wave (SPW1) amplitude and phase, and amplitude and phases of DS0, DW1, DW2 extracted from Fourier spectra computed using a window of ± 30 days over the equatorial region

during winter peaking at around 10 hPa/30 km. More interesting is the peak of 1-K amplitude of DW1 at the stratopause at 1 hPa/50 km and the absence of any significant amplitudes in the DS0 and DW2 components in the mid-latitude atmosphere. Figure 3 shows the mean temperature, SPW1 amplitude and phase for Fourier window sizes of ± 10 and ± 2 days. It can be seen that the sudden stratospheric warming [20] is observed more clearly during the winter of 2009/10 and there is more detail in the variation of SPW1 amplitude, which varies at a 60-day periodicity. During winter as the warming episode progresses, the region of warm air descends in altitude and so is the SPW1 amplitude. Only a reduction on window size for computing the Fourier transform has provided this insight into the variation of SPW1 amplitude.

The importance of this window size is two-fold. Firstly, it helps us in understanding the fine structure variations and the short-term variability in the wave amplitudes as described above. When these amplitudes are compared with results from other studies based on satellite data, like temperatures from COSMIC and TIMED, these provide a framework for a better understanding. Secondly, it has an important role to play in studies involving aliasing, particularly the aliasing of SPW1 and DS0/DW2. Many studies rely on satellite sampling of model data like CMAM30 to investigate aliasing in satellite data [2]. In the current study, we sampled the CMAM30 model in the Fourier space at non-grid longitudes and times, as if being sampled randomly by a satellite. The inverse Fourier transform was used to retrieve the temperature

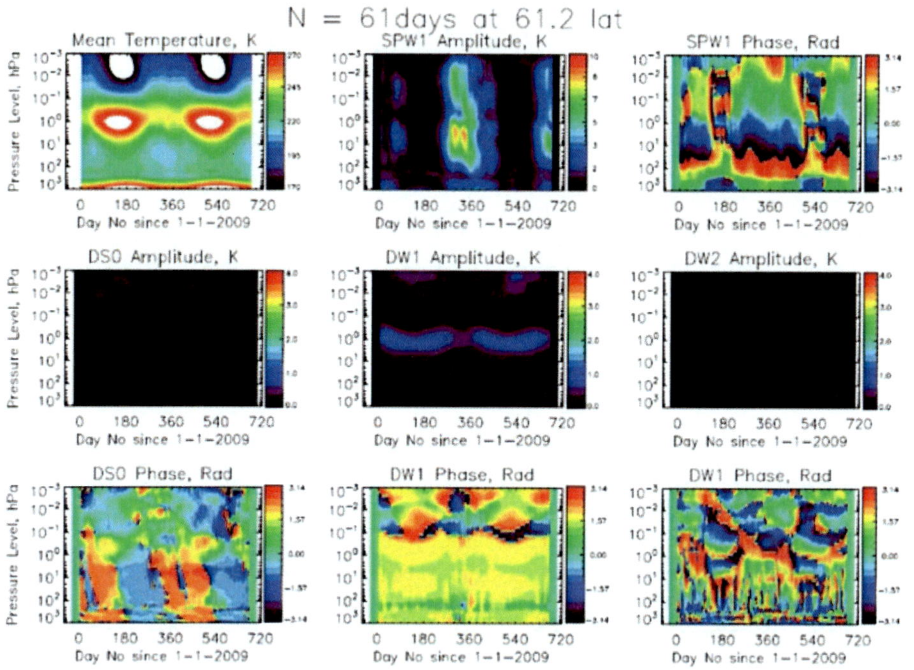

Fig. 2 Same as in Fig. 1, but for mid-latitudes (~60°N) using ± 30 days window

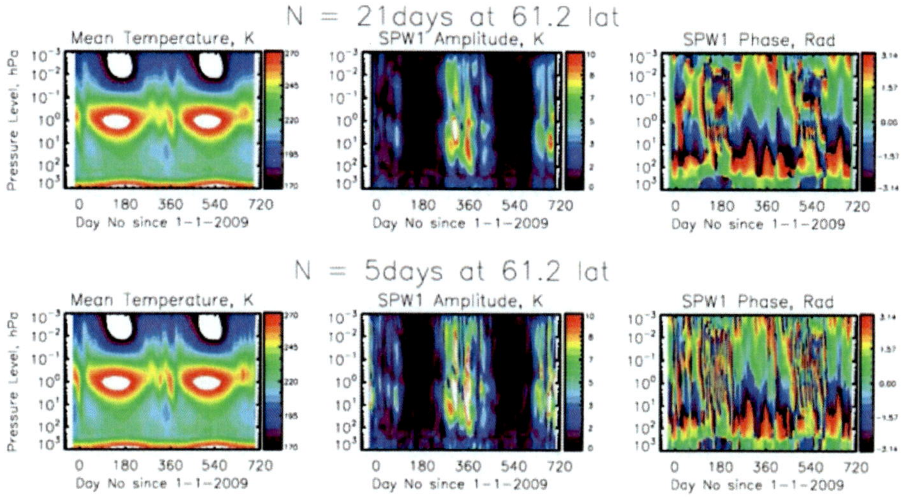

Fig. 3 Mean temperature and SPW1 amplitude and phase using ±10 and ±2 days over mid-latitudes (~60°N)

profiles at those longitudes and times. Usually, such subsampling is done using linear interpolation. We compare these two methods here.

Temperature profiles over the equator at two longitudes Lon[ix] and Lon[ix + 1] are considered and are shown in black and gray in Fig. 4 (Left panel). At the mid-point between these two longitudes, a temperature profile is estimated using linear interpolation (red profile) and also by sampling the Fourier space and then using the inverse Fourier transform to construct the temperature profile (blue profile). The right panel of the figure shows the differences between the temperatures at the grid points and those computed at the mid-point by sampling Fourier space and by interpolation. It can be seen that by using the method of interpolation the new profile at the midpoint longitude is also exactly in between the profiles (the differences are symmetrical about zero). However, by sampling the Fourier space, the new profile is not exactly in the middle. There are small but significant differences, particularly at the higher altitudes where there is significant geophysical variability. This is because this process considers the entire window of data by including variations over all frequencies and wave numbers in that window, while the former used only the nearest two neighbors. Hence, this process offers a better subsampling of the given model data. It may be noted here that the size of the window does not have any significant effect on the Fourier space sampling. By increasing the window size, we will be generating a Fourier space that contains the lower frequencies. It is observed in the current study that the results obtained by sampling the Fourier space obtained by using window sizes of ±2, ±10 and ±30 days have all given the same results. This study thus provides a framework to be able to investigate model data and satellite sampling to understand the aliasing problems involved in data analysis [2].

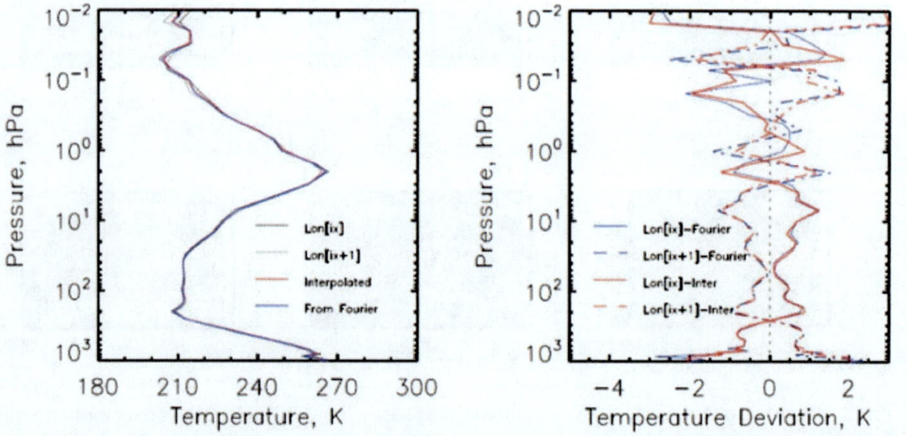

Fig. 4 (Left) Temperature profiles over the equator from two longitudes Lon[ix] and Lon[ix + 1] are shown in black and gray. Red profile shows the linearly interpolated values at the mid-point between Lon[ix] and Lon[ix + 1], and blue profile gives the temperature values obtained from sampling the Fourier space (and inverting). (Right) The differences between the temperatures at the grid points and those computed at the mid-point by sampling Fourier space and by interpolation

4 Conclusion

CMAM30 model temperature data is analyzed in this study using the two-dimensional fast Fourier transform using window sizes of ±2, ±10 and ±30 days. Wave and tidal components are extracted from the Fourier spectra and it is found that short-term variability is obtained by reducing the window size. Particularly, it is found that the SPW1 component varies at a periodicity of 60 days that was not identified earlier. It is also shown that sampling the Fourier space is a better way of subsampling the model data, rather than performing linear interpolation. However, the window size used in generating the Fourier space has little effect on the subsampling and subsequent inverting. It is thus concluded that large and non-overlapping windows are suited for generating a suitable Fourier space from model data. This has applications in sampling model data for satellite (like COSMIC and TIMED) locations and times to understand aliasing effects in the analysis of those datasets.

Acknowledgements This work is supported by Science and Engineering Research Board (SERB), Govt. of India, through grant ECR/2017/002258. Authors thank the Canadian Space Agency for providing free access to CMAM30 data.

References

1. Gan, Q., Du, J., Ward, W.E., Beagley, S.R., Fomichev, V.I., Zhang, S.: Climatology of the diurnal tides from eCMAM30 (1979 to 2010) and its comparison with SABER. Earth, Planets and Space **66**, 103 (2014)
2. Sakazaki, T., Fujiwara, M., Zang, X., Hagan, M.E., Forbes, J.M.: Diurnal tides from the troposphere to lower mesosphere as deduced from TIMED/SABER satellite data and six global reanalysis data sets. J. Geophys. Res. **117**, D13108 (2012)
3. Frobes, J.M.: Atmospheric Tides. 1. Model description and results for the solar diurnal component. J Geophys Res.(1982) 5222–5240 doi:https://doi.org/10.1029/Ja087ia07p05222.
4. Hagan, M.E.: Comparative effects of migrating solar sources on tidal signature in middle and upper atmosphere. J Geophys Res-Atmos. **101**(D16), 21213–21222 (1996)
5. Hagan, M.E., Frobes, J. M.: Migrating and nonmigrating diurnal tides in the middle and upper atmosphere excited by tropospheric latent heat release. J Geophys Res-Atmos. Vol. 107 (D24) (2002) doi:https://doi.org/10.1029/2001jd001236.
6. Chapman, S., Lindzen, R.S.: Atmosphere tides: Thermal and Gravitational. Gordon and Breach, New York (1970)
7. Andrews, D.G., Holton, J.R., Leovy, C.B.: Middle Atmosphere Dynamics. Elsevier 1st edn.(1987) pp. 5–6.
8. Du, J., Ward, W.E., Cooper, F.C.: The character of polar tidal signatures in the extended Canadian Middle Atmosphere Model (eCMAM). J Geophys Res-Atmos. **119**, 5928–5948 (2014)
9. Gan, Q., Zhang, S.D., Yi, F.: TIMED/SABER observation of lower mesospheric inversion layers at low and middle latitudes. J. Geo. Physics. Res- Atmos. Vol. 117 (2012) doi:https://doi.org/10.1029/2012jd017455.
10. Xu, J.Y., Smith A.K., Liu, M, Liu, X., Gao, H., Jiang, G.Y., Yuan, W.: Evidence for non-migrating tides produced by the interaction between tides and stationary planetary waves in

the stratosphere and lower mesosphere. J Geophys Res-Atmos. Vol.119 (2014) doi:https://doi.org/10.1002/2013JD020150.

11. Scinocca, J.F., McFarlane, N.A., Lazare, M., Li, J., Plummer, D.: The CCCma third generation AGCM and its extension into the middle atmosphere Atmos. Chem. Phys. **8**, 7055–7074 (2008)
12. de Grandpre, J., Beagley, S.R., Fomichev, V.I., Griffioen, E., McConnell, J.C., Medvedev, A.S., Shepherd, T.G.: Ozone climatology using interactive chemistry: Results from the Canadian Middle Atmosphere Model. Geophys. Res. **105**, 26475–26491 (2000)
13. McLandress, C., Plummer, D.A., Shepherd, T.G.: Technical Note: A simple procedure for correcting temporal discontinuities in ERA-Interim stratospheric temperatures for use in nudged chemistry-climate model simulations. Atmos. Chem. Phys, Discuss (2013)
14. Fomichev V.I., Ward, W.E., Beagley, S.R., McLandress, C., McConnell, J.C., McFarlane, N.A., Shepherd, T.G.: Extended Canadian Middle Atmosphere Model: zonal-mean climatology and physical parameterizations. J GeoPhys Res-Atmos. Vol.107 (D10) (2002) doi:https://doi.org/10.1029/200jd000479.
15. Beagley, S.R., Boone, C.D., Fomichev, V.I., Jin, J.J., Semeniuk, K., McConnell, J.C., Bernath, P.F.: First multi-year occultation observations of CO2 in the MLT by ACE satellite: observatons and analysis using the extended CMAM. Atmos Chem Phys **10**(3), 1133–1153 (2010). https://doi.org/10.5194/acp-10-1133-2010
16. Ward, W.E., Fomichev, V.I., Beagley, S.: Nonmigrating tides in equinox temperature fields from the Extended Canadian Middle Atmosphere Model(CMAM). Geophys Res Lett. 32(3) (2005) doi:https://doi.org/10.1029/2004gl021466.
17. Ekanayake, E.M.P., Aso, T., Miyahara, S.: Background Wind Effect on Propagation of non migrating diurnal tide in middle atmosphere. J. Atmos. Solar Terr. Phys. **59**, 401–429 (1997)
18. Forbes, J.M., Wu, D.: Solar tides as revealed by measurement of mesosphere temperature by the MLS experiment on UARS. J. Atmos. Sci **63**(7), 1776–1797 (2006). https://doi.org/10.1175/Jas3724.1
19. Smith, A.K.: Global Dynamics of the MLT. Surv Geophys **33**(6), 1177–1230 (2012)
20. Limpasuvan, V., Richter, J.H., Orsolini, Y.J., Stordal, F., Kvissel, O.K.: The roles of planetary and gravity waves during a major stratospheric sudden warming as characterized by WACCM. J. Atmos. Solar-Terr. Phys. **78–79**, 84–98 (2012)

Design and Analysis of an Efficient Queue Scheduling Scheme for Heterogeneous Traffics in BWA Networks

Tanusree Dutta, Prasun Chowdhury, Santanu Mondal, and Rabindranath Ghosh

Abstract The most important aspect of Broadband Wireless Access (BWA) networks is to provide heterogeneous traffic flows with a guarantee of Quality of Service (QoS). To this end, we have designed an efficient adaptive Weighted Hybrid Queue Scheduling Scheme (WHQSS) to support differential QoS requirements for traffic classes, which are heterogeneous in BWA networks. At first, the performance of WHQSS is analyzed in comparison with two important Queue Scheduling Schemes (QSSs), namely Priority Queue Scheduling Scheme (PQSS) and Hybrid Queue Scheduling Scheme (HQSS) those are designed *with* some major existing scheduling algorithms. A single analytical platform has been created using 3D Continuous Time Markov Chain (CTMC) model to investigate the performance in terms of several parameters like mean number of packets waiting in the queue, throughput, mean queuing delay, packet loss probability, and fairness index of the said QSSs. The comparative performance results clearly reveal the superiority of WHQSS providing optimal QoS guarantee and highest fairness than other QSSs for heterogeneous traffic classes.

Keywords BWA networks · Markov chain · Queue scheduling

T. Dutta (✉) · P. Chowdhury · R. Ghosh
St. Thomas' College of Engineering & Technology, Kolkata, India
e-mail: momtanu@gmail.com

P. Chowdhury
e-mail: prasun.jucal@gmail.com

R. Ghosh
e-mail: rnghosh@gmail.com

S. Mondal
Institute of Radio Physics & Electronics, University of Calcutta, Kolkata, India
e-mail: santanumondal2008@rediffmail.com

J. K. Mandal et al. (eds.), *Computational Intelligence and Machine Learning*, Advances in Intelligent Systems and Computing 1276,
https://doi.org/10.1007/978-981-15-8610-1_14

1 Introduction

In the recent past, there is an enormous interest in wireless communication systems with the demand for multimedia communication irrespective of the users' location. It has resulted in increased demands for heterogeneous services such as real-time video streaming and video conferencing, voice over IP, online gaming, and web browsing. This has led to shrinkage of the available bandwidth. Thus, the need for Broadband Wireless Access (BWA) networks are also increasing [1]. BWA is becoming a more challenging competitor to the conventional wired technologies like Digital Subscriber Line (DSL), Cable Modem, and even fiber optic cables due to last-mile wireless access [2]. Quality of Service (QoS) requirements is different for different services in heterogeneous media, which may refer to a different level of the desired performance parameters such as throughput, delay, and packet losses. As a result, there is always a significant challenge in designing a QoS-aware BWA network [3, 4].

Queue scheduler plays a major role in QoS aware BWA networks [5, 6] for managing heterogeneous applications. A queue scheduler serves the enqueued resource request of various applications, both real time (RT) and non-real time (NRT), ensuring the negotiated service-level agreements. It first decides the order of servicing the users request then manages the service queues to provide differentiated QoS guarantees to multiple users and multiple media flows [7]. The type of users in a network determines the Queue Scheduling Schemes (QSS) of the network. In the multi-class network, the queue scheduler categorizes the users into some predefined classes. Priority of each user is assigned according to its QoS requirements. Afterward, bandwidth is distributed among the users as per their priority in such a way so that fairness between the users is maintained. Hence, traffic differentiation and proportional fairness in resource management are, thus, the crucial features on which the research efforts are concentrated for BWA network to provide cost-effective last-mile communications.

Since the aim of this paper is to find the best Queue scheduling scheme for 5G BWA networks in general, the QoS services are mapped into three major QoS classes as given in Table 1. The classification is done based on the QoS need of the BWA services. Class-1, Class-2, and Class-3 represent three different requirements for QoS. Class-1 is considered to support the need for a fixed chunk of bandwidth to satisfy stringent delay for generating fixed-size data. Class-2 supports the need for

Table 1 BWA service types differentiation

QoS class	QoS need	BWA services
Class-1	Fixed chunk of Bandwidth to satisfy stringent delay	VoIP, Conversational voice and video
Class-2	Guaranteed Delay along with throughput and packet loss	Streaming video services
Class-3	Guaranteed Throughput and packet loss	FFTP, e-mail, and web browsing

guaranteed delay, throughput and packet loss to generate variable size data, while Class-3 supports the need for only guaranteed throughput and packet loss to perform tasks related to delay-insensitive data. Thus, all service types are mapped in terms of three different QoS as given in Table 1.

A. **Background of the Work**

In this section, some major existing scheduling algorithms mostly used for BWA networks to support differentiated QoS for 4G BWA networks are discussed. This is required to understand the relative performance of these algorithms in relation to a different class of services over 5G Platform. A two-tier Dynamic QoS-based bandwidth allocation (DQBA) scheme has been proposed in [8] to support delay and bandwidth requirements of heterogeneous service classes. In this hybrid scheduler, EDF and WFQ are used for VoIP and video streaming, respectively, whereas simple RR has been suggested for NRT class in tier 1 for intra-class scheduling. On the other hand, in tier 2, dynamic allocation of PQ and WFQ has been proposed for inter-class scheduling (heterogeneous traffic) to maximize the system capacity by efficiently utilizing its resources and thereby increases fairness. Here, the authors have used the traffic arrival rate to evaluate the priority factor as well as the weight factor. Simulation results show that although the proposed inter-class scheduling framework guarantees for delay-sensitive RT traffic, but it has low overall throughput for NRT traffic leading to the possibility of starvation under heavy traffic arrival of higher priority traffic. Moreover, no analytical model has been developed to justify the results. Hence, a significant challenge lies in defining the priority factor and weight factor for inter-class traffic scheduling which should ensure the service-level agreement of all traffic classes.

Authors in [9] have performed mathematical analysis for hybrids PQ-WFQ inter-class scheduler using Markov Chain Model. The mathematical model has been used for the evaluation of the effect of the Queue Weight Ratio (QWR) on various performance parameters like delay, throughput, packet loss, and fairness. But defining the proper weights and ways of packets to be served from each queue become the important factors for determining the overall system performance. There is a possibility of loosely bound on service guarantees without a proper tuning of weight values. Another key issue, which is not addressed in this paper, is the proper setting of the queue size, which is dynamically changed to meet the service requirement.

In [10], the authors have proposed a queue size-aware scheme for bandwidth allocation and rate control mechanisms. The Markov-Modulated Poisson Process is used to model the heterogeneous traffic classes to investigate the steady and transient states of the system. The proposed scheme has the main purpose of achieving higher throughput with smaller queue size as well as providing lower queuing delays. But the effect of packet drops is not considered while designing the system with a smaller queue size, which makes the scheme less efficient. The loss of packets in the queue may cause QoS degradation in the wireless domain, especially for the delay-sensitive RT traffic which needs guaranteed QoS support.

In order to solve the utilization optimization problem author in [11] has proposed Batch Arrival Queuing (BAQ) scheme for dynamically controlling queue size. He has developed a batch arrival model to assess QoS-aware packet scheduling performance for delay-sensitive RT traffic to maximize throughput and BU of the network by optimizing queue size with specified queuing delay and packet loss. But the problems remain for not considering valid packet classifier for QoS differentiation in BWA network. Moreover, the author has not mentioned any scheduling algorithm by which packets belonging to different traffic classes have been served. The BAQ model is only used to solve the utilization optimization problem for delay-sensitive RT traffic without considering the effect of NRT traffic classes.

B. **Approach of the Present Work**

At first, the interclass scheduling framework for BWA network to support heterogeneous traffic classes are designed in three major strategies, viz. PQSS, HQSS, and WHQSS. The literature review of scheduling algorithms leads us to design the interclass schedulers in the above three basic ways in which some major scheduling algorithms are actually implemented to provide differentiated QoS guarantee. Hence, our main contribution in this paper is to analyze the interclass QSSs, which are designed with some major existing scheduling algorithm for BWA network. The remainder of the paper is organized as follows. The design description of QSSs is given in section II. Section III provides a detailed description of the analytical model for QSSs using CTMC. Section IV sets the parameter for performance evaluation. Numerical results are shown in section V. Finally, Section VI concludes the paper.

2 Queue Scheduler Design

Three separate single-server queuing system for PQSS, HQSS, and WHQSS have been considered as shown in Fig. 1a, b, c, respectively. Each QSS comprised of three separate queues of finite length QL_1, QL_2, and QL_3. These are used to hold three types of traffic classes, Class-1, Class-2, and Class-3, respectively. The packets for three traffic classes is Poisson arrive with average rates of λ_1, λ_2, and λ_3, respectively. The service rate of each server is μ, and the inter-service time is exponentially distributed, with mean $1/\mu$. The number of packets present in the Class-1 traffic, Class-2 traffic, and Class-3 traffic, respectively, at any instant of time is represented by (L_1, L_2, L_3) [1].

As per the QoS requirement of the users, Class-1 traffic should require hard QoS support and have. That is why the three queuing server systems considered providing a guaranteed priority highest priority among all other traffic classes while getting service from the queuing server system to the Class-1 traffic over the others. Packets in Class-1 traffic are always scheduled with service rate μ. Packets in Class-2 traffic and Class-3 traffics are only scheduled in the absence of Class-1 packets. QSS algorithms are given in Table 2.

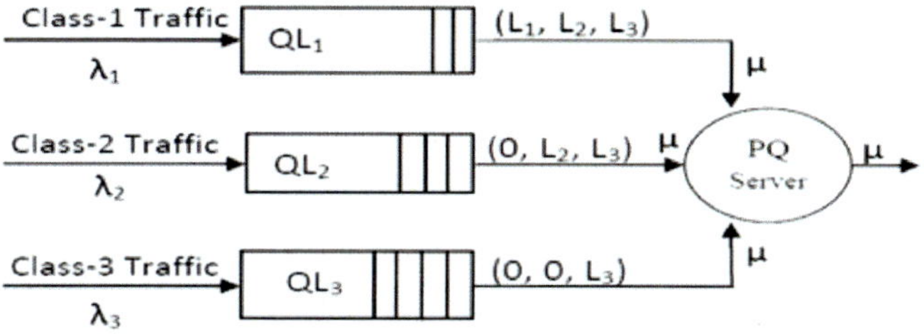

(a) Priority Queue scheduling scheme (PQSS)

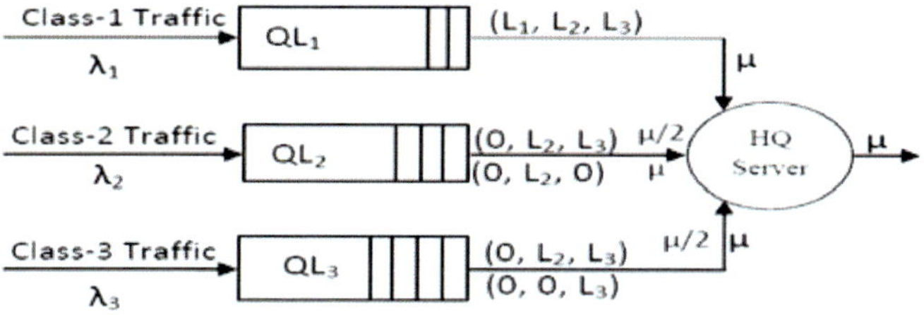

(b) Hybrid Queue scheduling Scheme (HQSS)

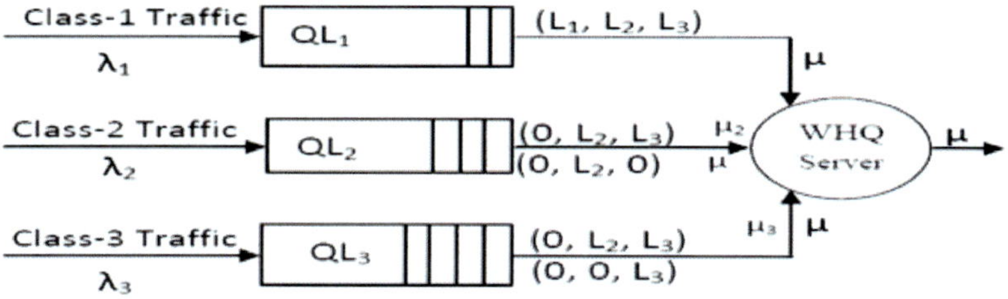

(c) Weighted Hybrid Queue scheduling scheme (WHQSS)

Fig. 1 Queue Scheduling Schemes (QSSs)

The service rates μ_2 and μ_3 mentioned in the WHQSS algorithm are dynamically changed based on the QWR to meet the service guarantee. Class-2 traffic corresponding to the RT video traffic is assigned a higher weight w_2.

The service rates μ_2 and μ_3 mentioned in the WHQSS algorithm are dynamically changed based on the QWR to meet the service guarantee. Class-2 traffic corresponding to the RT video traffic is assigned a higher weight w_2. The delay insensitive Class-3 traffic is assigned with a lower weight w_3. Thus, Class-2 and Class-3 traffics will achieve an average service rate μ_2 and μ_3, respectively, as given in (1) and (2).

$$\mu_2 = \frac{\mu w_3}{w_2 + w_3} \tag{1}$$

Table 2 QSS algorithms

PQSS Algorithm: For each service request in PQSS
if ($L_1 = 0$ and $L_2 = 0$)
then serve packets from QL_3 with service rate μ
else if ($L_1 = 0$)
then serve packets from QL_2 with service rate μ
else
serve packets from QL_1 with service rate μ
End for
	HQSS Algorithm: For each service request in HQSS
	If ($L_1 = 0$) && ($L_2 \neq 0$) && ($L_3 \neq 0$)
	then serve packets from QL_2 and QL_3 in RR fashion with service rate $\mu/2$ each
	else if ($L_1 = 0$) && ($L_2 \neq 0$)
	then serve packets from QL_2 with service rate μ
	else if ($L_1 = 0$) && ($L_3 \neq 0$)
	then serve packets from QL_3 with service rate μ
	else
	serve packets from QL_1 with service rate μ
	End for
		WHQSS Algorithm: For each service request in WHQSS
		If ($L_1 = 0$) && ($L_2 \neq 0$) && ($L_3 \neq 0$)
		then serve packets from QL_2 and QL_3 using WFQ with service rate μ_2 and μ_3, respectively
		else if ($L_1 = 0$) && ($L_2 \neq 0$)
		then serve packets from QL_2 with service rate μ
		else if ($L_1 = 0$) && ($L_3 \neq 0$)
		then serve packets from QL_3 with service rate μ
		else
		serve packets from QL_1 with service rate μ
		End for

$$\mu_3 = \frac{\mu w_3}{w_2 + w_3} \tag{2}$$

3 Analytical Model

In this paper, the performance evaluation of the three separate single-server queuing system for PQSS, HQSS, and WHQSS have been performed in a single analytical platform using 3D CTMC Model [12] as shown in Fig. 2. In this platform, the three QSSs differ only in their state transition rate for scheduling as given in Table 3. A continuous change in its current state occurs in the queue scheduling server due to the occurrence of events (i.e., incoming and outgoing of packets). For a more accurate analysis of its performance, it is necessary to observe the short-lived states of the system. This is possible only if the queue scheduling server uses CTMC modeling. Furthermore, the CTMC model does not increase the complexity of the queue scheduling server since it has only three dimensions.

The queue scheduling server [13] receives the bandwidth requests from the queue of three traffic classes directly. The server changes state from one to another upon the incoming to or outgoing from a queue corresponding to a particular traffic class. So, the next state of the queue scheduling server depends only on the present state but does not depend on the previous states. Therefore, the states of the queue scheduling server form a Markov Chain as shown in Fig. 2. Hence, the individual QSS can be analytically modeled using the state transition diagram subject to the corresponding state transition rate as depicted in Table 3.

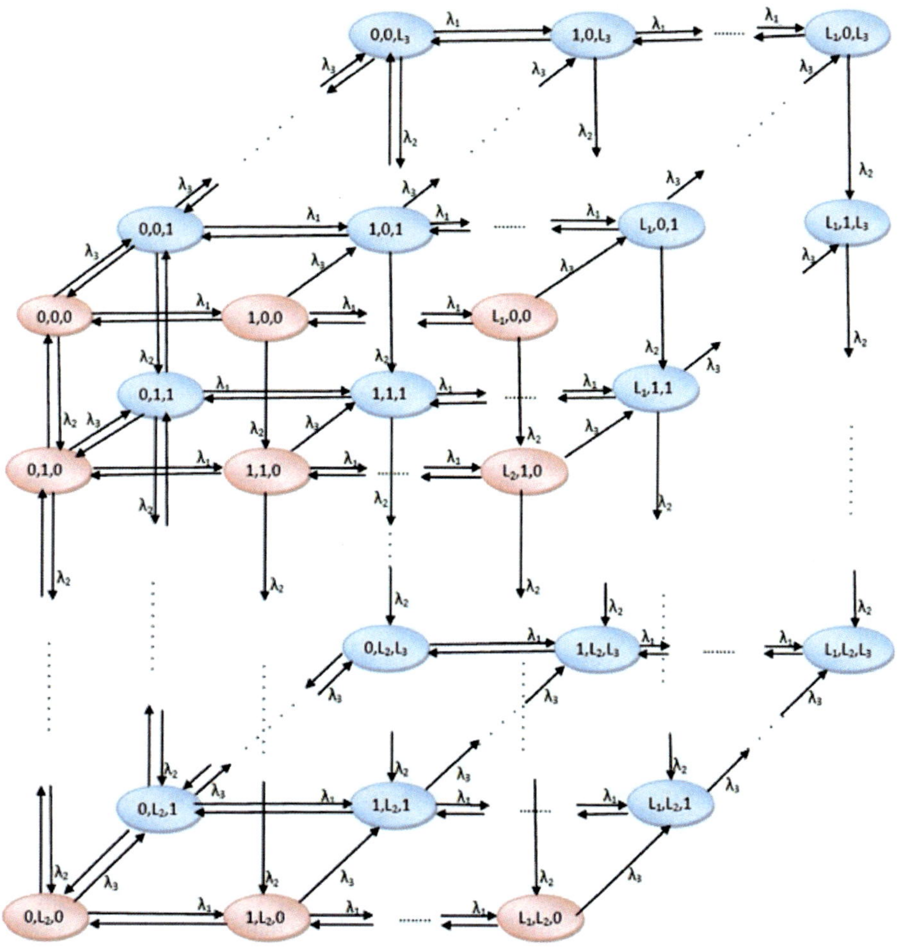

Fig. 2 A generalized 3D CTMC model for all QSSs

In the 3D Markov Chain, state S = (L_1, L_2, L_3) represents that the server currently has L_1, L_2, and L_3 number of Class-1 packets, Class-2 packets, and Class-3 packets, respectively. As per Poisson distribution, the incoming packets have the rate of arrival as λ_1, λ_2. and λ_3, respectively. $1/\mu$ is the mean of the exponential distribution of the service times of Class-1 traffic, Class-2 traffic, and Class-3 traffic. The state-space '**S**' for single-server queuing system is obtained based on Eq. (3).

$$S = \{s = (L_1, L_2, L_3) | (L_1 \leq QL_1) \wedge (L_2 \leq QL_2) \wedge L_3 \leq QL_3\} \quad (3)$$

The transition may take place from state (L_1, L_2, L_3) to ($L_1 + 1$, L_2, L_3), from state (L_1, L_2, L_3) to (L_1, $L_2 + 1$, L_3), and from state (L_1, L_2, L_3) to (L_1, L_2, $L_3 +$ 1). This implies that a packet from Class-1, Class-2, and Class-3 traffics enters into

Table 3 State transition rate for scheduling different traffic classes

Scheduling scheme	Traffic type	State transitions		Service rate for scheduling
		Present state	Next state	
PQSS	Class-1	(L_1, L_2, L_3)	$(L_1\text{-}1, L_2, L_3)$	μ
	Class-2	$(0, L_2, L_3)$	$(0, L_2\text{-}1, L_3)$	μ
	Class-3	$(0, 0, L_3)$	$(0, 0, L_3\text{-}1)$	μ
HQSS	Class-1	(L_1, L_2, L_3)	$(L_1\text{-}1, L_2, L_3)$	μ
	Class-2	$(0,L_2, L_3)$	$(0, L_2\text{-}1, L_3)$	$\mu/2$
		$(0, L_2, 0)$	$(0, L_2\text{-}1, 0)$	μ
	Class-3	$(0, L_2, L_3)$	$(0, L_2, L_3\text{-}1)$	$\mu/2$
		$(0, 0, L_3)$	$(0, 0, L_3\text{-}1)$	μ
WHQSS	Class-1	(L_1, L_2, L_3)	$(L_1\text{-}1, L_2, L_3)$	μ
	Class-2	$(0, L_2, L_3)$	$(0, L_2\text{-}1, L_3)$	μ_2
		$(0, L_2, 0)$	$(0, L_2\text{-}1, 0)$	μ
	Class-3	$(0, L_2, L_3)$	$(0, L_2, L_3\text{-}1)$	μ_3
		$(0, 0, L_3)$	$(0, 0, L_3\text{-}1)$	μ

the queue with rates λ_1, λ_2, and λ_3, respectively. Moreover, the state transition rate from state (L_1, L_2, L_3) to (L_1-1, L_2, L_3) is equal to the Class-1 traffic service rate μ, which is the outgoing rate of a packet from the queue of Class-1 traffic.

The state transition rate for scheduling different traffic classes are summarized in Table 3 assuming $(L_1 > 0, L_2 > 0$ and $L_3 > 0)$.

Let the steady state probability of the state $s = (L_1, L_2, L_3)$ be represented by $\pi_{(L_1,L_2,L_3)}(s)$. The Markov chain is irreducible and it is observed that the outgoing and incoming states for a given state 's', the steady-state probabilities of all states of the single-server queuing system have been evaluated with the help of the normalized condition that sum of all steady-state probabilities equals to one [1]. That is,

$$\sum_{s \in S} \pi_{(L_1,L_2,L_3)}(s) = 1 \tag{4}$$

π_{L_i} is the marginal state probability and it gives the probability that L packets of Class i $(where, i \in 1, 2, 3)$ are present in the queue. It can also be calculated [8] based on the steady-state probability $\pi_{(L_1,L_2,L_3)}(s) = 1$.

$$\pi_{L_1} = \sum_{L_3=0}^{QL_3} \sum_{L_2=0}^{QL_2} \pi_{(L_1,L_2,L_3)} \tag{5}$$

$$\pi_{L_2} = \sum_{L_3=0}^{QL_3} \sum_{L_1=0}^{QL_1} \pi_{(L_1,L_2,L_3)} \tag{6}$$

$$\pi_{L_3} = \sum_{L_1=0}^{QL_1} \sum_{L_2=0}^{QL_2} \pi_{(L_1,L_2,L_3)} \tag{7}$$

It is possible to derive the following QoS performance parameters of the system from the marginal state probabilities.

A. **Mean marginal number of packets (mmnp) waiting in the queue**

$$mmnp_i = \sum_{L_i=0}^{QL_i} (L_i * \pi_{L_i}) \tag{8}$$

B. **Mean marginal throughput (mmt)**
Throughput can be defined as the average rate at which packets go through the system in the steady state. In steady-state condition, the rate of flow of incoming traffic and outgoing traffic are in equilibrium. The product of the arrival rate of a particular class of traffic i and the marginal state probability $\pi_{\mathrm{L_i}}(where, i \in 1, 2, 3)$ is the mean marginal throughput (mmt). Hence, mmt_i is given by

$$mmt_i = \lambda_i * \sum_{L_i=0}^{QL_i-1} \pi_{L_i} \tag{9}$$

C. **Mean marginal queuing delay (mmqd)**
As per Little's Law [12], *mmqd* is the ratio of mean marginal number of packets in a class of queue to the throughput of that corresponding class of queue. Hence, $\mathrm{mmqd_i}$ of a particular class i of queue is given by

$$mmqd_i = \frac{mmnp_i}{mmt_i} \tag{10}$$

D. **Marginal packet loss probability (mplp)**
When the queue of a particular traffic class is full, an incoming packet can be dropped. The marginal packet loss probability is equal to the marginal state probability when queue is completely full of packets. Thus, $mplp_i$ for a particular traffic class i is given by

$$mplp_i = \pi_{QL_i} \tag{11}$$

Table 4 Values of the system parameters

Parameters	Values
Traffic Arrival rate ratio of Class-1, Class-2, and Class-3 traffics (λ_1:λ_2:λ_3)	3:2:1 and 5:4:1
Mean service rate (μ)	20 packets/sec
Variable Packet size	64–1024 bytes
Queue weight ratio (w_2:w_3)	7:3
SDQ of Class-2 traffic	20–200 ms
Parameters	Values
Traffic Arrival rate ratio of Class-1, Class-2, and Class-3 traffics (λ1:λ2:λ3)	3:2:1 and 5:4:1
Mean service rate (μ)	20 packets/sec
Variable Packet size	64–1024 bytes
Queue weight ratio (w2:w3)	7:3
SDQ of Class-2 traffic	20–200 ms

4 Settings of the System Parameters

Table 4 summarizes the values of the system parameters used in the analytical model for the evaluation of the performance of the PQSS, HQSS, and WHQSS. The traffic arrival rate ratio denoted by TARR of Class-1, -2, and -3 traffics is 3:2:1. This is because of the practical scenario in which the network experiences the highest arrival request from traffic Class-1. It receives the lowest arrival request from traffic Class-3. The performance analysis of Class-1, -2, and -3 traffics are also verified with a higher TARR of 5:4:1 to observe the performance patterns of each class with increased traffics.

5 Performance Analysis

The different performance parameters for the three QSSs are analyzed with designed CTMC models by writing programs in MATLAB platform under various QoS constraints and then the comparative analysis is made to understand the relative performance of the considered QSSs. Justifications behind all the numerical results have also been provided.

A. **Comparative performance of PQSS, HQSS, WHQSS**

Figures 3, 4, 5, 6 show the comparative performances of PQSS, HQSS, and WHQSS on various QoS parameters like packets waiting in the queue, throughput, queuing delay, and packet loss. These performances are observed for Class-1, Class-2, and Class-3 traffics against the TARR 3:2:1 and 5:4:1. It is observed that there is no variation in the performance of PQSS, HQSS, and WHQSS (Figs. 3a, 4a, 5a, and 6a)

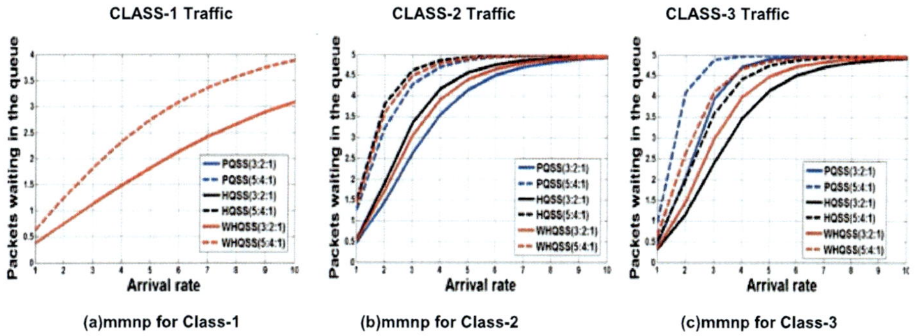

Fig. 3 Mean marginal number of packet (mmnp) waiting in the queue

for traffic Class-1 under a TARR. As the priority of the Class-1 traffic is maintained highest for all scheduling schemes, the performance curves of PQSS, HQSS, and WHQSS are merged together revealing the similar QoS performance under a TARR. However, the QoS performances are different for different scheduling schemes in case of Class-2 and Class-3 traffics. A close observation of Figs. 3b, 4b, 5b, and 6b reveal that PQSS provides the best performance in terms of packets waiting in the queue, throughput, queuing delay, and packet loss for Class-2 traffic. The worst performance is observed for Class-2 traffic under HQSS. On the other hand, the scenario is reversed for Class-3 traffic as observed in Figs. 3c, 4c, 5c, and 6c.

In PQSS, Class-2 traffic gets the opportunity to serve its packet with a full bandwidth just after queue of Class-1 traffic becomes emptied, whereas in HQSS, Class-2 traffic has to share the bandwidth in RR fashion with the Class-3 traffic. This leads to the bandwidth degradation of the Class-2 traffic while serving packets from its queue resulting in poor QoS performance. Though PQSS provides the best performance to Class-2 traffic, Class-3 traffic has to suffice starvation as observed in Fig. 4c. On the other hand, though HQSS has eliminated the problem of starvation and boost up the performance of Class-3 traffic, Class-2 traffic has to suffice poor throughput, queuing delay, and packet loss as observed in Figs. 4b, 5b, and 6b, respectively. However, the proposed WHQSS has solved the above-mentioned problem by providing optimal QoS guarantee to both Class-2 and Class-3 traffic as observed in Figs. 3, 4, 5, 6.

6 Conclusion

In this paper, by categorizing the QoS requirement of BWA traffics into several classes, an inter-class scheduler was designed named as WHQSS. The analysis revealed that the optimal performance of WHQSS provides the best QoS guarantee for BWA networks when compared to other QSSs. It has reduced the delay of RT applications and guaranteed the throughput of NRT applications. WHQSS also ensured fairness in allocating resources among all connections appropriately. However, the

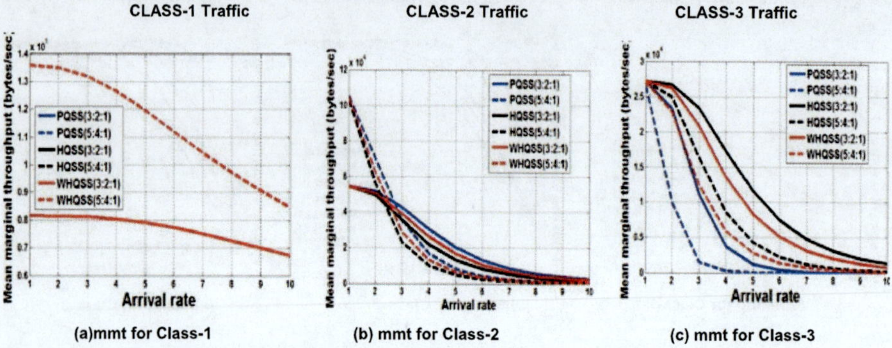

Fig. 4 Mean marginal throughput (mmt)

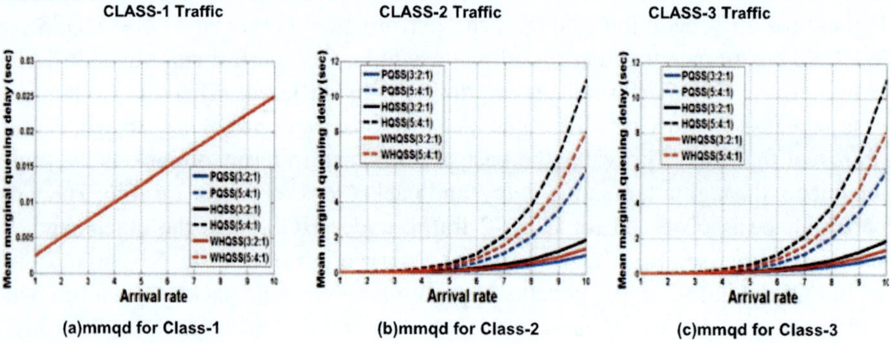

Fig. 5 Mean marginal throughput (mmt)

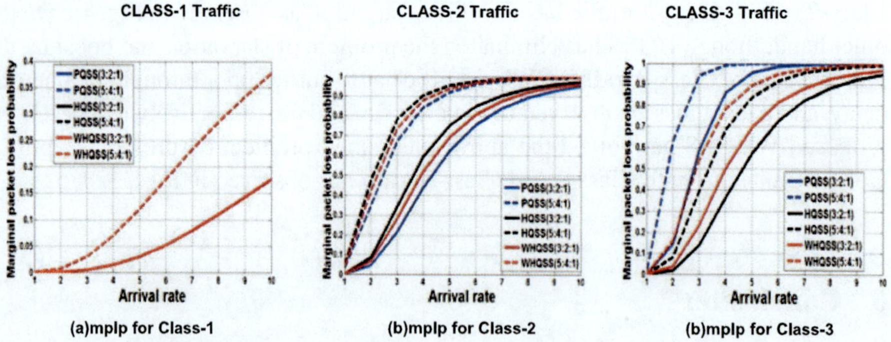

Fig. 6 Mean marginal queuing delay (mmqd)

development of channel-aware WHQSS by introducing the PHY layer parameters is our future work.

References

1. Chowdhury, P., Misra, I.S.: Queue size analysis of QoS-aware Weighted Hybrid Packet Scheduling Scheme for BWA networks. In: 19th IEEE Interntional Conference on Networks (ICON) (2013)
2. Wen, Y.F., Lin, F.Y.S., Tzeng, Y.C., Lee, C.T.: Backhaul and routing assignments with end-to-end qos constraints for wireless mesh networks. Wirel. Personal Commun. **53**(2), 211–233 (2010)
3. LAN/MAN Standards Committee of the IEEE Computer Society and the IEEE Microwave Theory and Techniques Society, IEEE Standard for Air Interface for Broadband Wireless Access Systems , IEEE Std 802.16TM-2017.
4. Ghosh, A., Rat.asuk, R., Mondal, B., Mangalvedhe, N., Thomas, T.: LTE-Advanced: next-generation wireless broadband technology. IEEE Wirel. Commun. **17**(3), 10–22 (2010)
5. Oad, A., Subramaniam, S.K., Zukarnain, Z.A.: Enhanced uplink scheduling algorithm for efficient resource management in IEEE 802.16. EURASIP J. Wirel. Commun. Netw. **2015**(Article number: 3) (2015)
6. Khalil Shahid, M., Shoulian, T., Shan, A.: Mobile broadband: comparison of mobile wimax and cellular 3G/3G+ technologies. Inf. Technol. J. **7**(4), 570–579 (2008)
7. Kurose, J., Ross, K.: Computer Networking: A Top-down Approach, 7th edn. Pearson (2017)
8. Esmailpour, A., Nasser, N.: Dynamic QoS-based bandwidth allocation framework for broadband wireless networks. IEEE Trans. Veh. Technol. **60**(6) (2011)
9. Wang, L., Min, G., Kouvatsos, D., Jin, X.: An analytical model for hybrid PQ-WFQ scheduling scheme for WiMAX netwotks. In: Proceedings of International Conference on Wireless VITAE, pp. 492–498 (2009)
10. Niyato, D., Hossain, E.: Queue-aware uplink bandwidth allocation and rate control for polling service in IEEE 802.16 broadband wireless networks. IEEE Trans. Mob. Comput. **5**(6), 668–679 (2006)
11. Chen, C.: Dynamis classified buffer control for QoS-aware packet scheduling in IEEE 802.16/WiMAX networks. IEEE Commun. Lett. **14**(9), 815–817 (2010)
12. Ross, S.M.: Introduction to Probability Models, 10th edn. University of Southern California Los Angels, California (2010)
13. Muhajir, A., Binatari, N.: Queuing system analysis of multi server model at XYZ Insurance Company in Tasikmalaya city. AIP Conf. Proc. **1886**, 040004 (2017)

Simulation of Cardiac Action Potential with Deterministic and Stochastic Pacing Protocols

Ursa Maity, Anindita Ganguly, and Aparajita Sengupta

Abstract Cardiac arrhythmia is a major group of heart diseases that are caused by irregular heartbeats and often leads to sudden deaths. This work analyses a detailed numerical model of the mammalian ventricular cell. The aim is to provide a simplified yet comprehensive simulation model for understanding the effects of ionic concentration, basic cycle length (BCL) on the generation of action potential and employing pacing protocols with ease. Variation in these inputs as well as in the pacing results in significant changes in the membrane action potential duration (APD). These results are clearly demonstrated without delving into the physiological basis of the same. This approach shall be helpful in modifying any cell model to investigate the clinical relevance of various pacing protocols that can be employed in the future. The Luo–Rudy Phase I model of the mammalian ventricular cell has been used for simulations to demonstrate the effects of the rate dependence of activation in cardiac tissues, referred to as the Wenckebach periodicity.

Keywords Cardiac arrhythmia · Action potential · Sudden death · Electrophysiology · Deterministic pacing · Stochastic pacing

1 Introduction

In the last few decades, there has been significant progress in the field of cardiac physiology, especially in the application of pacing protocols in mathematical models

U. Maity (✉) · A. Sengupta
Department of Electrical Engineering, Indian Institute of Engineering Science and Technology (IIEST), Shibpur, Howrah, India
e-mail: ursa5597@student.ubc.ca

A. Sengupta
e-mail: sgaparajita@gmail.com

A. Ganguly
TARE Faculty, IIEST and Guru Nanak Institute of Technology (GNIT), Kolkata, India
e-mail: aninditaganguly80@gmail.com

J. K. Mandal et al. (eds.), *Computational Intelligence and Machine Learning*, Advances in Intelligent Systems and Computing 1276,
https://doi.org/10.1007/978-981-15-8610-1_16

of the human heart. Many models based on mammalian ventricular action potential have been developed over the course of time. Hodgkin and Huxley et al. [7, 8] developed a model that represents the electrical behaviour of the membrane modelled using ionic currents and capacitance of membrane. McAllister et al. [15] modelled the electrical activity in Purkinje fibres which are specialized conducting fibres present in cardiac myocytes. A mathematical model was developed by Beeler and Reuter [1] using the framework of Hodgkin–Huxley equations and also applying voltage-clamp techniques. The Beeler–Reuter model was then modified into the Luo–Rudy Phase I model in 1991 (Luo et al.) and improved into phase II model in 1994 (Luo et al.).

Othman et al. [17] carried out a hardware implementation of the Luo–Rudy Phase I model. The basic framework of the action potential in cardiac myocytes still holds and has been used as the basis of the Simulink Model drawn in MATLAB for this work. The same model has also been used in a previously unpublished work by the authors. In this paper, additional simulations have been carried out on the aforementioned model using pacing protocols.

The aim of the current work is to exhibit the effects of pacing and stimulus variations in the Luo–Rudy Phase I model in a purely mathematical framework. Less focus is given on the physiological details keeping in mind the queries of the less trained in the biological aspect of the topic. Importance has been given to parameters which can be varied directly to obtain variation in outputs. The corresponding changes in the action potential generation are recorded using Simulink Data Inspector.

The results obtained are consistent with the experimental findings as well as theoretical simulations in the existing literature. The effect of change in stimulus duration is the first step in studying the various protocols and analysing the results for diagnosis and treatment using pacing protocols such as deterministic and stochastic pacing (Dvir et al. 2013 and 2014). Different APD restitution (APDR) curves can be obtained by different pacing protocols such as oscillatory, random and linear (Wu et al. 2004).

2 Methods

2.1 Theory and Mathematical Formulation

The membrane voltage (V_m) depends on the sum of the ionic channel currents (i_{total}) and the stimulus current (i_{stim}) and the membrane capacitance (C_m) as given in (1). This is the basic equation by Hodgkin, Huxley et al.

$$\frac{dV_m}{dt} = \frac{-(i_{total} + i_{stim})}{C_m} \tag{1}$$

The current i_{total} is represented as follows in (2) (Luo–Rudy 1991)

$$i_{total} = i_{Na} + i_{Ca} + i_K + i_{Ki} + i_{plat} + i_{bg} \tag{2}$$

where i_{Na} is the fast sodium current, i_{ca} is the slow inward calcium current, i_K is the time-dependent potassium current, i_{Ki} time-independent potassium current, i_{plat} is the plateau potassium current and i_{bg} is the background current.

All the ionic currents are determined by a set of gating variables and corresponding voltage rate-dependent constants. The formulations for the respective currents (Hodgkin, Huxley et al., Luo–Rudy, Beeler–Reuter) are shown below in (3)–(8), respectively.

$$i_{Na} = g_{Na} \cdot m^3 \cdot h \cdot j \cdot (V_m - E_{Na}) \tag{3}$$

$$i_{Ca} = g_{Ca} \cdot d \cdot f \cdot (V_m - E_{Ca}) \tag{4}$$

$$i_K = g_K \cdot x \cdot x_{inac} \cdot (V_m - E_K) \tag{5}$$

$$i_{Ki} = g_{Ki} \cdot k_{inac} \cdot (V_m - E_{Ki}) \tag{6}$$

$$i_{plat} = g_{plat} \cdot k_{plat} \cdot \left(V_m - E_{plat}\right) \tag{7}$$

$$i_{bg} = 0.03921 \cdot (V_m + 59.87) \tag{8}$$

The rest of the dynamics are defined as follows in (9)–(15):

$$E_{Na} = \frac{RT}{F} \ln\left(\frac{[Na_o]}{[Na_i]}\right) \tag{9}$$

$$E_{Ca} = 7.7 - 13.0287 \ln[Ca_i] \tag{10}$$

$$E_K = \frac{RT}{F} \ln\left(\frac{[K_o] + pr_{NaK}[Na_o]}{[K_i] + pr_{NaK}[Na_i]}\right) \tag{11}$$

$$E_{Ki} = E_{plat} = \frac{RT}{F} \ln\left(\frac{[K_o]}{[K_i]}\right) \tag{12}$$

$$g_K = 0.282\sqrt{\frac{[K_0]}{5.4}} \tag{13}$$

$$g_{Ki} = 0.6047\sqrt{\frac{[K_0]}{5.4}} \tag{14}$$

$$\frac{d[Ca_i]}{dt} = 0.07[Ca_i] - 10^{-4}(i_{Ca} - 0.07) \tag{15}$$

Table 1 Description of the gate variables

Gate variable	Function
m	Activation for i_{Na}
h	slow inactivation for i_{Na}
j	fast inactivation for i_{Na}
x	Activation for i_K
x_{inac}	İnactivation for i_K
k_{inac}	İnactivation for i_K
d	Activation for i_{Ca}
f	İnactivation for i_{Ca}

Here, g_{ion} is the maximum conductance of the corresponding ionic channel ($mScm^{-2}$), E_{ion} is the reversal potential of the corresponding ion (mV), $[E_o]$ and $[E_i]$ are the extracellular and intracellular concentrations of ion E (mM), respectively and k_{plat} is the inactivation gate for i_{Kplat}. The value of g_{Na} is taken as $23mScm^{-2}$ (Beeler–Reuter model), g_{si} as $0.09mScm^{-2}$ and g_{Kplat} as $0.0183\ mScm^{-2}$. Also, pr_{NaK} is the permeability ratio of Na/K and its value is taken as 0.01833. R is the gas constant with value 8.314 $JK^{-1}\ mol^{-1}$, T is the ambient temperature in K and F is Faraday's constant with value $9.6485 \times 10^4\ C\ mol^{-1}$.

The variables m, h, j, x, d and f are the respective gating variables for the corresponding ionic channels. Their values range between 0 and 1. The details are listed in Table 1.

The gate variables are computed using an ordinary differential equation (ODE) of the form shown in (16),

$$\frac{d\gamma}{dt} = \theta_\gamma - \gamma\left(\theta_\gamma + \mu_\gamma\right) \tag{16}$$

Here, the ODE in (16) holds for any gate variable γ and having rate constants θ_γ and μ_γ (ms^{-1}). The rate constants have been calculated using the equations of the Luo–Rudy Phase I model.

2.2 *Simulink Model*

The Simulink model is made using MATLAB functions and blocks (Maity et al.). It has been observed that change of time interval between two stimulation pulses has a major effect on the wave pattern of Action Potential generation. The basic cycle length was changed and the APD curves were obtained for 260 ms, 350 ms, 500 ms, 630 ms, 750 ms and 890 ms, respectively. The distinguishable change in the APD patterns is attributed to the rate-dependent activation failure, also referred to as Wenckebach periodicity by Luo and Rudy. The pacing protocols have also been discussed in the upcoming sections. The length of each stimulation interval is

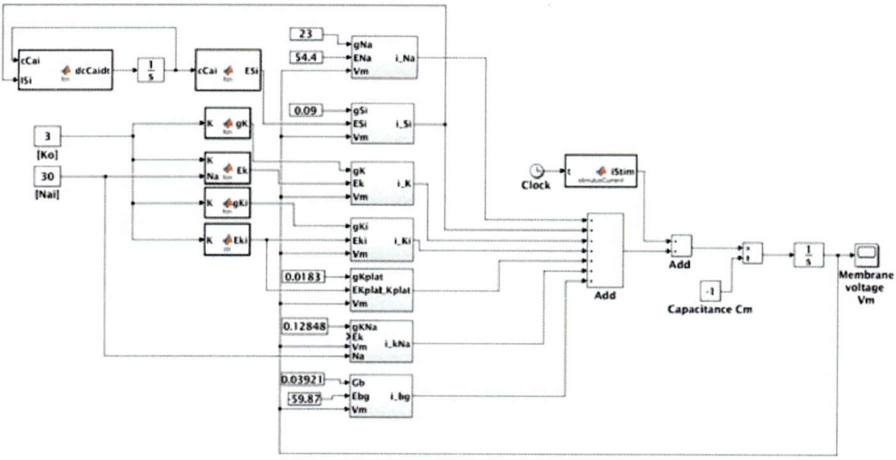

Fig. 1 Simulink Model of Luo–Rudy Phase I model

characteristic of the type of protocol being employed. Stimulus intensity is set at 80 μAcm^{-2}, and the duration of pulse applied is 1 ms.

A schematic representation of the Simulink model created using the equations in Sect. 2.1 is given in Fig. 1.

3 Results

3.1 *Periodic Rate Dependence*

The simulation results obtained on varying the stimulus intervals from 890 ms to 260 ms are observed. The stimulus interval directly determines the basic cycle length of the APD curves. It is observed that as the BCL value is reduced, the APD wave patterns become irregular. This causes a change in stimulus-to-response (S:R) ratio as demonstrated by Luo–Rudy. The larger the value of basic cycle length, the lesser is the S:R ratio. This is due to the Wenckebach phenomenon which is popularly referred to as the second-degree atrioventricular block (Mobitz I and II). The first type, Mobitz I, results in heart skipping beats at regular intervals, which makes the body to cope with it. But, the second type, Mobitz II, consists of irregular patterns from which the body is unable to recover [3].

The relationship between the APD length and BCL is shown in Fig. 2. The average values of APD (ms) are plotted against the various BCL (ms) values applied.

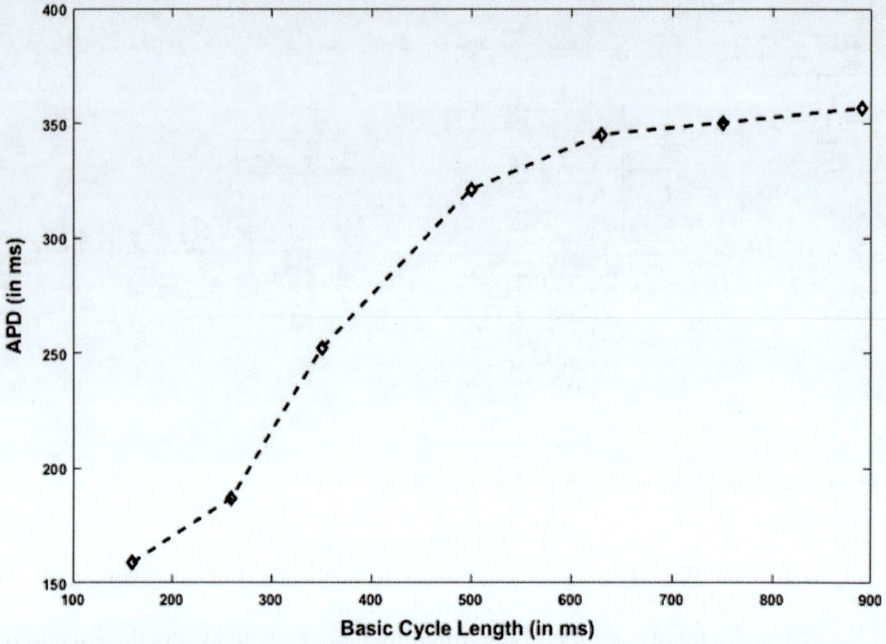

Fig. 2 Variation of APD lengths with Basic Cycle Length (BCL)

3.2 Deterministic Pacing and Stochastic Pacing

The dynamic pacing protocol employed in our Simulink model follows the mathematical concept of expressing the total cycle length (CL) in ms, as a summation of basic cycle length (BCL) and some noise. In case of deterministic pacing, this noise is almost negligible, i.e. the intervals between consecutive stimulus pulses are constant. Dvir et al. (2013–2014) represented the noise as Gaussian white noise, G(0, σ) with zero mean and varying standard deviation σ. In case of deterministic pacing, the value of standard deviation is set to zero.

$$CL = BCL + G(0, \sigma) \tag{17}$$

Choi et al. [2] and Lemay et al. [10] suggested a third-order autoregressive model express the APD of the (n + 1)th beat as a function of the APDs and DIs of the three previous beats.

In case of deterministic pacing, the relation between diastolic interval (DI) and BCL and APD can be given as follows in (18):

$$BCL = APD + DI \tag{18}$$

In case of stochastic pacing, (17) and (18) is combined to give the following relation as given in (19):

$$CL = APD + DI + G(0, \sigma) \tag{19}$$

Both the deterministic and stochastic pacing protocols have been applied to the Simulink model. The results with and without stochastic variations have been shown in Fig. 12. For deterministic pacing, the stimulus interval was set at 500 ms and a stimulus current of magnitude 80 μAcm^{-2} was applied.

In order to apply stochastic variations in the protocol, a Gaussian white noise of zero mean and a standard deviation of 10 ms is applied using the MATLAB function *'randn'* which was added to the regularly spaced intervals using the MATLAB function *'cumsum'*. This is added to the stimulus intervals stored in an array with values 0 ms, 500 ms, 1000 ms and so on. As observed in Fig. 3, the wave patterns are distinguishable for the two pacing protocols employed.

The physiological relevance of this result is validated by Dvir et al. (2013) with the fact that the cardiac tissue is activated better in a stochastic rhythm of pacing compared to a deterministic rhythm. Usually, a low Heart Rate Variability (HRV) which is directly related to low standard deviation is useful in the prediction of arrhythmia in case of lethal cardiac events. In this study, a basic model of random pacing with Gaussian variations has been applied and its role in ventricular electrical activity has been observed. This method is one of the initial steps in developing more advanced and complex programming sequences for prediction of arrhythmia, as well as for the achievement of cardiac immunity by eliminating arrhythmogenic

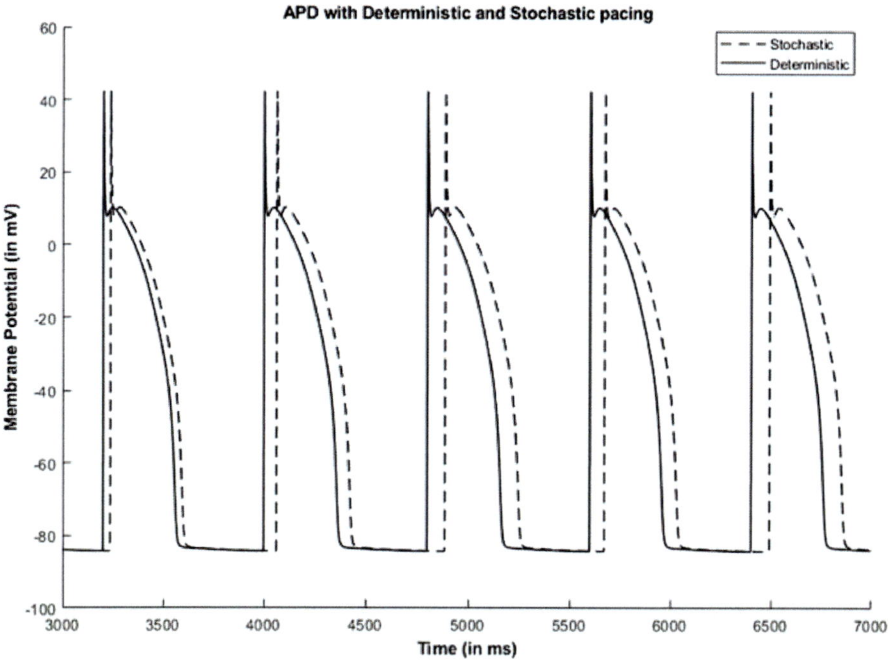

Fig. 3 APD curves with deterministic pacing (solid line) and stochastic pacing (dashed line)

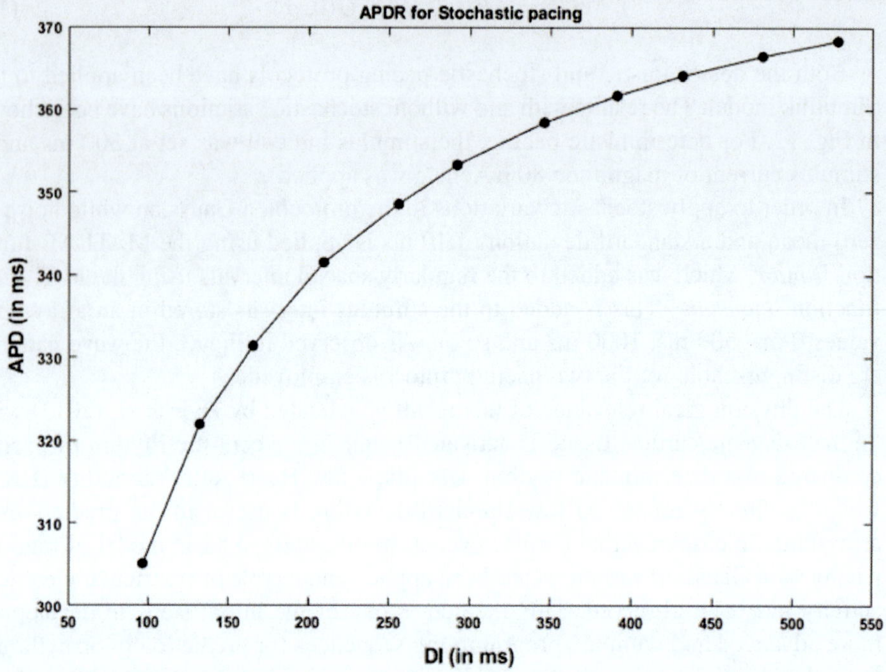

Fig. 4 APD versus DI over a BCL range of 400–900 ms with the application of stochastic pacing

factors. Dvir et al. (2013) mentioned that such factors include APD restitution slope and APD alternans propensity. The APDR curve (APD versus DI) when stochastic pacing protocol is applied is shown in Fig. 4. Basic cycle length was varied between 400 and 900 ms to obtain the APD and DI values.

4 Conclusion

This study is based on a detailed Simulink model for the action potential of the mammalian ventricular cell. Pacing protocols play an important role in diagnosis as a risk predictor for arrhythmia. It is observed that low stimulus intervals cause irregularities in the action potential curves indicating a periodic rate dependence of the action potential. This study shows how a stochastic variation in the pacing protocol affects the APD. A future attempt may be made to distinguish and adapt these protocols in order to rectify different categories of cardiac arrhythmia.

Acknowledgements This work is the outcome of a SERB, GoI sponsored TARE Project and is partially sponsored by the Department of Electrical Engineering, IIEST, Shibpur. The authors are grateful to Prof. Jan Kucera, Department of Physiology, University of Bern, for exposing them to further investigation on the Luo–Rudy model and pacing protocols.

References

1. Beeler, G.W., Reuter, H.: Reconstruction of the action potential of ventricular myocardial fibres. J Physiol (Lond) **268**, 177–210 (1977)
2. Choi, B.R., Liu, T., Salama, G.: Adaptation of cardiac action potential durations to stimulation history with random diastolic intervals. J. Cardiovasc. Electrophysiol. **15**, 1188–1197 (2004)
3. Clifford Gari D., Azuaje, F., McSharry Patrick, E.: Advanced methods and tools for ECG data analysis. Artech House (2006)
4. Dvir, H., Zlochiver, S.: Stochastic cardiac pacing increases ventricular electrical stability—a computational study. Biophys. J. **105**, 533–542 (2013)
5. Dvir, H., Zlochiver, S.: The Interrelations among Stochastic Pacing, Stability, and Memory in the Heart. Biophys. J. **107**, 1023–1034
6. Faber, G.M.: and Rudy Y, Action Potential and Contractility Changes in $[Na^{+}]_i$ Overloaded Cardiac Myocytes: A Simulation Study. Biophys. J . **78**, 2392–2404 (May)
7. Hodgkın, A. L., Huxley, A. F.: The components of membrane conductance in the giant axon of Loligo. J. Phy8iol. **1I6**, 473–496 (1952a)
8. Hodgkin, A.L., Huxley, A.F.: A quantitative description of membrane current and its application to conduction and excitation in nerve. J Physiol (Lond) **117**, 500–544 (1952b)
9. Kameyama, M., Kakei, M., Sato, R., Shibasaki, T., Matsuda, H., Irisawa, H.: Intracellular Na^{+} activates a K^{+} channel in mammalian cardiac cells. Nature **309**, 354–356 (1984)
10. Lemay, M., de Lange, E., Kucera, J.P.: Uncovering the dynamics of cardiac systems using stochastic pacing and frequency domain analyses. PLOS Comput. Biol. e1002399 (2012)
11. Levi, A.J., Dalton, G.R., Hancox, J.C., Mitcheson, J.S., Issberner, J., Bates, J.A., Evans, S.J., Howarth, F.C., Hobai, I.A., Jones, J.V.: Role of intracellular sodium overload in the genesis of cardiac arrhythmias. J. Cardiovasc. Electrophysiol. **8**, 700–721 (1997)
12. Luo, C.H., Rudy, Y.: A model of the ventricular cardiac action potential. Depolarization, repolarization, and their interaction. Circ. Res. **68**(6), 1501–1526 (1991)
13. Luo, C.H., Rudy, Y.: A dynamic model of the cardiac ventricular action potential. I. Simulations of ionic currents and concentration changes. Circ. Res. **74**, 1071–1096 (1994)
14. Maity U., Ganguly A., Sengupta A., Simulation of Action Potential Duration and its dependence on $[K]_o$ and $[Na]_i$ in the Luo Rudy phase I model, (accepted), COMSYS 13–15 (2020)
15. McAllister, R.E., Noble, D., Tsien, R.W.: Reconstruction of the electrical activity of cardiac Purkinje fibres. J Physiol (Lond) **251**, 1–59 (1975)
16. Nolan, J., Batin, P.D., Fox, K.A.: Prospective study of heart rate variability and mortality in chronic heart failure: results of the United Kingdom heart failure evaluation and assessment of risk trial (UK-heart). Circulation **98**, 1510–1516 (1998)
17. Othman, N., Jabbar, M.H., Mahamad, A.K., Mahmud, F., Rudy, L.: Phase I excitation modeling towards HDL Coder Implementation for real-time simulation. 978–1–4799–4653–2/14/© 2014 IEEE
18. Williams, L.: Wilkins, ECG Interpretation Made Incredibly Easy (Incredibly Easy! Series®), LWW, 6th edn. (2015)

A Study on Development of PKL Power

K. A. Khan, Md. Afzol Hossain, Salman Rahman Rasel, M. Ohiduzzaman, Farhana Yesmin, Lovelu Hassan, M. Abu Salek, and S. M. Zian Reza

Abstract The PKL electricity is included under biomass electricity. The scientific name of PKL is Bryophillum pinnatum Leaf. PKL means Pathor Kuchi Leaf, which is the local name in Bangladesh. It is a very innovative technology in the present world. It has been developed the electricity from PKL in Bangladesh first. That is why sometimes it is called "Bangla Electricity" by somebody. In the present study, it has been found the open-circuit voltage (Voc), short-circuit current (Isc), maximum power (P_{max}), voltage drop for different loads, consuming voltage, consuming current

K. A. Khan (✉)
Department of Physics, Jagannath University, Dhaka 1100, Bangladesh
e-mail: kakhan01@yahoo.com

Md. A. Hossain
Department of Chemistry, University of Dhaka, Dhaka 1100, Bangladesh
e-mail: afzal.chemistry@gmail.com

S. R. Rasel
Local Government Engineering Department (LGED), Fulbaria, Mymensing, Bangladesh
e-mail: salman_rasel80@yahoo.com

M. Ohiduzzaman
Department of Physics, Jashore University of Science & Technology, Jashore 7408, Bangladesh
e-mail: ohid@just.edu.bd

F. Yesmin
Department of Civil Engineering, Dhaka Polytechnic Institute, Dhaka, Bangladesh
e-mail: farhanayesmin1975@gmail.com

L. Hassan
Department of Physics, Jahangirnagar University, Savar, Dhaka, Bangladesh
e-mail: lovelu.iu.bd@gmail.com

M. A. Salek
Department of Chemistry, Adamjee Cantonment College, Dhaka 1206, Bangladesh
e-mail: salek@acc.edu.bd

S. M. Z. Reza
Department of Physics, Uttara University, Dhaka, Bangladesh
e-mail: smzianreza@gmail.com

J. K. Mandal et al. (eds.), *Computational Intelligence and Machine Learning*, Advances in Intelligent Systems and Computing 1276,
https://doi.org/10.1007/978-981-15-8610-1_17

pattern for different loads of 1-KW PKL micro (Pathor Kuchi Leaf)-power system. The variation of load voltage (V_L), load current (I_L), and load power (P_L) with the variation of time have also been studied. Most of the results have been tabulated and graphically discussed.

Keywords Performance · 1-KW PKL power · Technology · Practical utilizations

1 Introduction

The PKL electricity is a very innovative technology across the globe. So that it is not so popular to society. Now, it is needed to develop for the betterment of the people for practical utilization. Though people are using solar electricity at the rural off-grid areas across the globe [1–19]. But they are facing numerous problems during night time and rainy seasons for the lack of solar radiation. To keep it in mind, it has been designed and developed as 1-KW PKL micro-power plant for practical utilization [20–37]. It works with same performance both the day and night time. Here, PKL extract is used as an electrolyte and Zn and Cu as electrodes [38–58]. To save the land space, it has also been designed and developed a stand, where the total electrodes, electrolytes, and the converter can be placed properly. This device is called a stand-alone system for PKL power production. It is mentioned that the total loads were also set up on that stand. The switchboard was also set up there. It is also very interesting that there are some by-products can be obtained from the research work [59–67]. The by-products are methane gas, zinc-based biofertilizer, and hydrogen gas. The loads were two street lights, one tube light, one ceiling fan, one table fan, two LED lights, and one LED indicator light, etc., which were also set up on the same stand (Fig. 4). A circuit board and one circuit breaker were also set up there on the same stand. It is very interesting to mention that the performance of PKL electricity is better in nighttime than daytime. The reason behind that the presence of the malic acid is more in the PKL at nighttime than daytime. It is known to all that the main compound in the PKL is weak organic acid like citric acid, isocitric acid, and malic acid [68–74]. The people who are living in the off-grid region and those who are not able to use the computer in this area, this work will help them to access the electricity for internet connections. People can access this electricity instead of solar photovoltaic electricity [75–82]. This work will be the guideline for electricity generation for practical utilization in near future.

2 Objectives

1. To design and fabricate 1-KW portable PKL micro-power plant for practical utilization.
2. To grow the consciousness about the PKL micro-electricity to society.

3 Methodology

This leaf is also called Miracle leaf. It grows everywhere. The main compound of this leaf is citric acid, isocitric acid, and malic acid which all are weak organic acid. Figure 1a, b, c, d shows a PKL tree, PKL cultivation, plucking of leaf, leaf prepared for extract, and then preparation of PKL extract [89], respectively.

Figure 2 shows the Extract of the PKL. It has been blended by a blender machine with a few percentages of water. Then it has been filtered by a filter machine. After filtration, PKL waste was obtained as a residue (Fig. 3) which is used as a biofertilizer [83–85].

We can again use this waste for making PKL extract mixed with water for electricity generation. Even we can use this waste until 12 months to prepare PKL extract for the same [86–88].

(a) PKL tree (b) PKL Collection

(c) PKL after Collection (d) Extract preparation

Fig. 1 PKL (*Bryophillum Pinnatum* Leaf) tree collection from the farmer and juice preparation

(a) PKL after collection (b) Extract preparation (c) Extract filtration

Fig. 2 Experimental preparation of PKL (*Bryophillum pinnatum* Leaf) and extract

(a) one time used (b) Two times used

Fig. 3 Experimental preparation of a PKL waste after getting PKL extracts

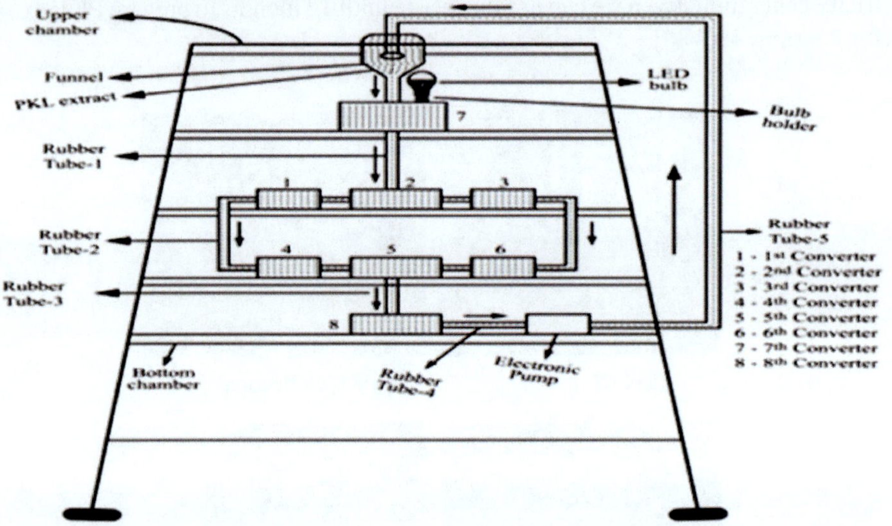

Fig. 4 Design of an experimental setup of a 1-KW PKL power production from the PKL extract

Figure 4 shows the design of an experimental setup of a 1-KW PKL Power Production from the PKL extract. It is shown that there are eight converters in this system. There are four chambers lower to top to set up PKL converters and the loads. There are five rubber tubes for connecting the converters with each other for flowing the PKL extract. Firstly, the extract was put into the funnel and the funnel was connected to the converter-7. Converter-7 was fixed with chamber-4. Converter-7 is a relatively small compared to the other converters. It was directly connected to converter-2 by a rubber tube-1. Converter-2 is connected with converter-1 and converter-3. Similarly, converter-1 and converter-3 are also connected to converter-4 and converter-6, respectively. Collector-5 stands between collector-4 and collector-6. Collector-8 stands on the upper surface of the first chamber. It was connected with an electric pump and the pump was also connected with converter-7 by a rubber tube. The whole device was assembled in a mechanical system based on the stand-alone system shown in Fig. 4.

Fig. 5 An experimental setup of a 1-KW PKL power production from the PKL extract

It was designed and fabricated six insulated boxes (Fig. 5) which were available in the local market of Bangladesh. The copper and zinc plates were set up in the boxes scientifically with parallel and series combinations. The Zn and Cu plates were parallel in connection in each box and the boxes are connected in series connections with each other for getting required current, voltages. and power. The boxes were set up in a stand-alone chamber. There was a small insulated box filled up with Zn and Cu plates including with LED lantern on the top of the chamber. There was also a small insulated box filled up with Zn and Cu plates under the bottom of the chamber.

Figure 6 shows the fabrication of the 1-KW PKL power production system. It is fabricated based on the design of the PKL Power system. Here, it is shown converter-1, converter-2, converter-3, converter-4, converter-5, converter-6, converter-7, and converter-8. It is also shown that there are two street lights = 600 W, two LED lights = 60 W, one ceiling fan = 80 W, one table fan = 30 W, one tube light = 40 W, and one self-operated electric pump = 50 W which carried PKL extract forcibly from converter-8 to converter-7 (red color). The light used at converter-7 as an indicator of the 1-KW PKL Power system. The overall 1-KW PKL power system works as the description (Fig. 6) of the earlier discussion. The total power used in the system

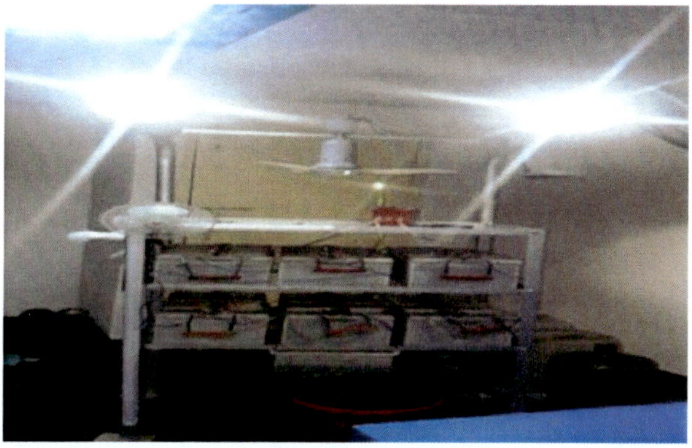

Fig. 6 Experimental setup of a 1-KW PKL power production from the PKL extract—front view

Fig. 7 Experimental setup of a 1-KW PKL power production from the PKL extract—left view

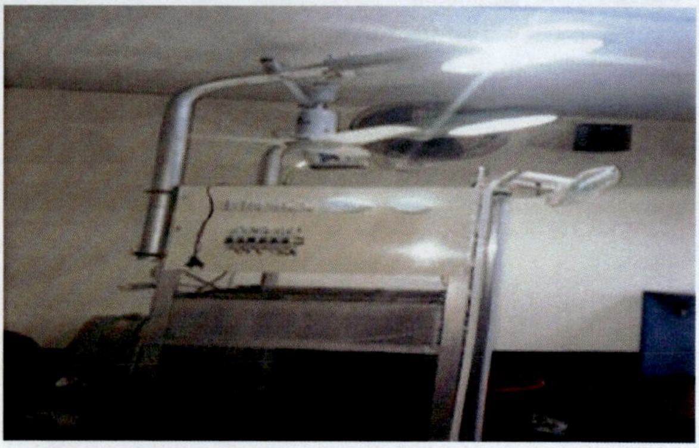

= 860 W, which is around 80% of the source. It is mentioned that the output power should be 80% of the source power.

It is shown (Fig. 7) the right side of the 1-KW PKL micro-power plant where there is a control board consists of a circuit breaker, some switches, and two LED lights. The total circuit of the 1-KW PKL micro-power plant is controlled by the switches and the circuit board.

4 Nernst Equation and PKL Cell

Josiah Willard Gibbs had formulated a theory to predict whether a chemical reaction is spontaneous based on the free energy,

$$\Delta G = \Delta G^0 + \mathrm{RT} \ln \mathrm{Q} \quad (1)$$

Here, ΔG is the change in Gibbs free energy, ΔG° is the cell potential when Q is equal to 1, T is the absolute temperature (Kelvin), R is the universal gas constant, and Q = Quotient constant. Based on Gibbs' work, Nernst extended the theory to include the contribution from electric potential on charged species [7, 48–50]. The change in Gibbs free energy for an electrochemical cell can be related to the cell potential. Thus, Gibbs' theory becomes,

$$\mathrm{nF}\Delta \mathrm{E} = \mathrm{nF}\Delta \mathrm{E}_0 - \mathrm{RT} \ln \mathrm{Q} \quad (2)$$

Here, n is the number of electrons/mole product, F is the Faraday constant (coulombs/mole), and ΔE is cell potential. Now, dividing Eq. (2) by the amount of charge transferred (nF) to arrive at a new equation which now bears his name, that is Nernst equation:

$$\Delta \mathrm{E} = \Delta \mathrm{E}_0 - \ln \mathrm{Q} \quad (3)$$

Assuming the standard state conditions (T = 25 °C) and R = 8.3145 J/ (K·mol), the equation above can be expressed on base-10 logarithm as shown below [46–48]:

$$\Delta E = \Delta E_0 - \log Q \tag{4}$$

where

$$[Zn^{2+}]/[Cu^{2+}][H^{+}] \tag{5}$$

5 Observations of the Concentration of Cu^{2+} Varying with Time

The presence of Cu^{2+} ion in PKL juice solution as a secondary salt acts as a reactant ion. Thus, the presence of Cu^{2+} ion increases both potential and current flow with time Cu^{2+} reduces to Cu and so the concentration of Cu^{2+} ion decreases [49–51].

6 Reactions

$$Cu^{2+}_{(aq)} + 2e^{-}_{(aq)} \rightarrow Cu_{(s)} \tag{6}$$

Again, the anode undergoes corrosion to give the product ion Zn^{2+} by the following the reaction:

$$Zn\ (s) - 2e^{-}(aq) \rightarrow Zn^{2+}(aq) \tag{7}$$

So, the variation of concentration of Zn^{2+} ion will be helpful to this study. But Zn^{2+} can't be determined by UV–Vis Spectrophotometer. For this reason, AAS has been used to determine this Zn^{2+}.

7 Results and Discussion

The results have been collected and tabulated carefully. Most of the results have been graphically represented.

Table 1 shows the number of cells in the six boxes was 72, where the maximum open-circuit voltage was 83.34 V. The short-circuit current was 12.89 A. The maximum produced power was 1074.25 W (Table 2).

Table 1 Current–voltage relations

Number of cell	Open-circuit Voltage, V_{OC} (Volt)	Short-circuit Current, I_{SC} (A)	Maximum Power, Pmax = (Voc)(Isc) (W)
72	83.34	12.89	1074.25

Table 2 Variation of load voltage with the variation of local time

Open-circuit Voltage, V_{OC} (Volt)	Local time (hrs)	Load Voltage (Volt)
83.34	00	50.28
	10	49.66
	15	49.24
	20	48.96
	25	48.75
	30	48.55
	35	48.52
	40	48.40
	45	48.20
	50	48.06

The total load power was tried to keep 80% of the total produced power (around 1000 W). From Fig. 8, it is shown that the load voltage gradually decreased almost linearly with the variation of local time. It decreased linearly from starting to 30 h and then it was remained same up to 36 h and then after it decreased up to 50 h. It is also shown (Table 2) that the total voltage variation was around 2.22 V.

Table 3 shows the consuming voltage with the variation of consuming time. From Fig. 9, it is shown that the consuming voltage varies with time almost linearly up to 50 h. It is also shown from Fig. 9 that the consuming voltage varies up to 15 h almost linearly and, at the 10 end hours, it varies almost linearly, but from 15 to 40 h, it increased almost exponentially.

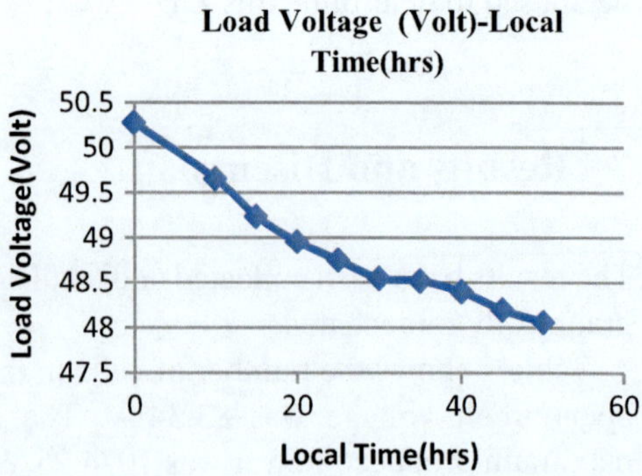

Fig. 8 Load voltage versus local time graph

Table 3 Consuming time versus consuming voltage table

Consuming time (hrs)	Consuming voltage (volt)
00	33.06
10	33.68
15	34.1
20	34.38
25	34.59
30	34.79
35	34.82
40	34.94
45	35.14
50	35.28

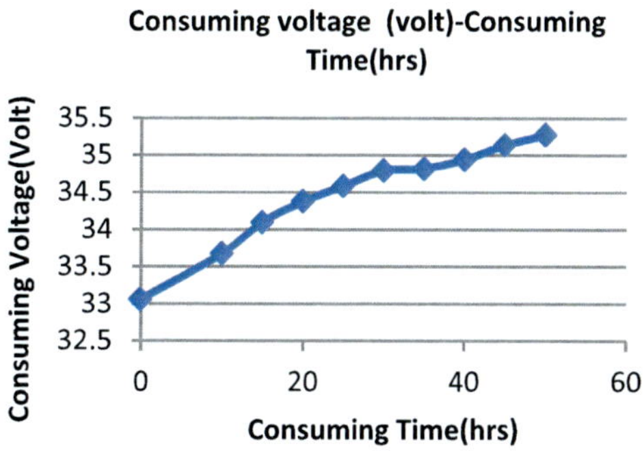

Fig. 9 Consuming voltage versus consuming time graph

It was collected minimum voltages for 4 days which was tabulated (Table 4) and graphically represented (Fig. 10). It is seen that the collected minimum load voltages were different for different days and it is also seen that it decreases for different days. It is also shown that it almost linearly decreases for the first day and then it decreases almost exponentially for next 3 days.

It was also collected the maximum voltage for 4 days (Table 5) and that was tabulated in Fig. 11. The maximum voltage decreases almost linearly for the first day and then it decreases almost exponentially for next 3 days.

Table 4 Table for minimum voltage with different loads

Date (Day)	Minimum voltage with Load (Volt)
Day-1	47.29
Day-2	46.32
Day-3	45.92
Day-4	43.75

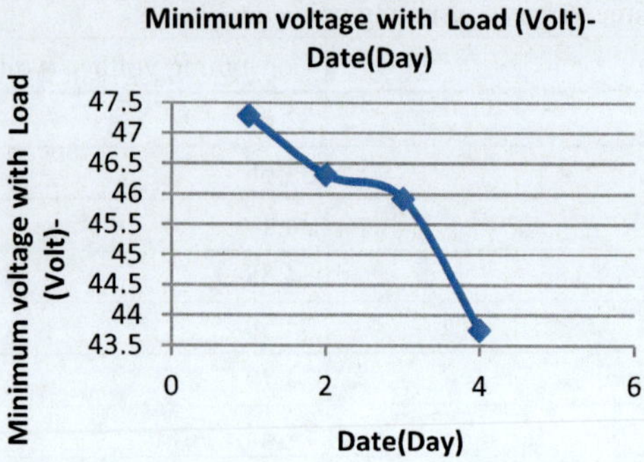

Fig. 10 Minimum voltage versus date graph

Table 5 Variation of maximum voltage of a PKL module for different loads

Date of the month of the year (Day)	Maximum Voltage with load (Volt)
Day-1	50.28
Day-2	48.96
Day-3	48.82
Day-4	48.26

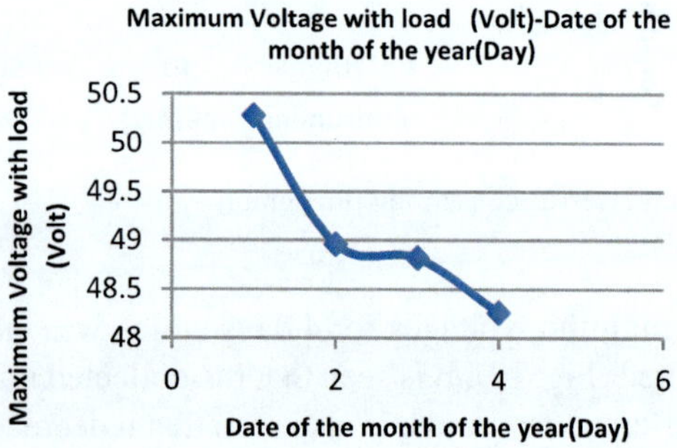

Fig. 11 Maximum voltage versus date graph

It was collected the voltage without load (Table 6) for 35 h to study the self-discharge characteristics. It is shown that the voltage without load was almost constant up to first 10 h and it decreased slightly only 0.06 V for next 5 h and then again it decreased to 0.03 V, 0.02 V, 0.05 V, and 0.06 V for every 5 h, respectively, up to 35 h. It is also shown that the decreasing rate increases with time. It is also shown that it was tolerable. But this desirable rate was kept constant and which is practically impossible. It is also shown that the total difference of this voltage is 0.24 V for 35 h (Table 7).

It is shown that the consuming voltage varies with time (Fig. 13). It is shown (Table 7) that the consuming voltage was almost constant with time for 35 h. The difference

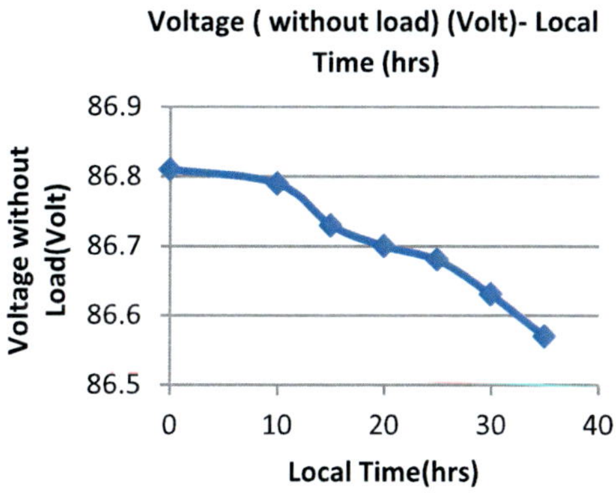

Fig. 12 Voltage without load versus local time graph

Table 6 Variation of voltage (Volt) with the variation of local time (hrs) for without load

Local time (hrs)	Voltage (without load) (Volt)
00	86.81
10	86.79
15	86.73
20	86.70
25	86.68
30	86.63
35	86.57

Table 7 Table for consumed voltage by PKL lantern with the variation of local time

Local time (hrs)	Consuming Voltage by the load (Volt)
00	30.09
10	31.73
15	30.76
20	29.58
25	28.82
30	27.08
35	26.11

of this voltage was around 5.62 V. The voltage variation was 1.6 V (increasing) for the first 10 h and then it was 1.07 V (decreasing) for next 5 h, then 1.18 V (decreasing) for next 5 h, 0.76 V (decreasing) for next 5 h, 1.74 V (decreasing) for next 5 h, and finally, it was 0.91 V (decreasing) for last 5 h.

Table 8 shows the load currents for different loads, whereas the open-circuit voltage was 82.2 V and the short-circuit current was 12.30 V. It is also shown that the current depends on the resistance. It is also shown that the open-circuit voltage for each cell was 1.10 V.

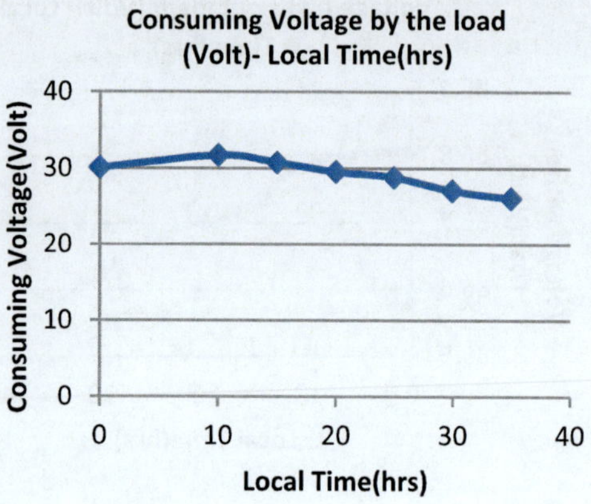

Fig. 13 Voltage without Load versus local time graph

Table 8 Current–voltage relationships with different loads

Number of cell	Open-circuit voltage, Voc (Volt)	Short-circuit current, Isc (A)	Load Current, I(A) for different load
72	82.2	12.30	2.05
			3.07
			4.81
			5.74
			6.83
			6.86
			10.60
			11.09
			11.33
			11.40
			11.14
			11.60
			11.70
			11.74
			11.76

Table 9 shows the load currents for different loads, whereas the open-circuit voltage was 82.2 V and the short-circuit current was 12.30 V. It is also shown that the current depends on the resistance. It is also shown that the open-circuit voltage for each cell was 1.10 V (Tables 10 and 11).

Figure 14 shows a study for 150 h. It is shown that the voltage decreases with local time but very slowly. The total voltage drop (Table 10) was around 1 V which was from 47.5 V to 48.5 V. It is shown that it decreases linearly for the first 10 h and then it increases linearly for next 5 h and then it was constant for 10 h and then after

Table 9 Current–voltage relationship with different loads

Number of cell	Open-circuit voltage, Voc (Volt)	Short-circuit current, Isc (A)	Load Current, I (A) for different loads
72	82.2	12.30	1.00
			1.33
			1.50
			1.60
			1.70
			1.71
			1.60
			1.64
			1.67
			1.70
			1.57
			1.60
			1.59
			1.55

it decreases linearly. In this way sometimes it decreases and sometimes increases. But the voltage variation was limited within 1 V.

The load voltage was taken by changing the load resistance (Fig. 15). It was also taken for 150 h. Here the total voltage variation (Table 11) was from 48.85 V to 49.66 V. The amount of this variation was around 0.81 V. The variation is almost linear but sometimes it looks fluctuations from 48.85 V to 49.66 V.

8 Conclusions

The long-term study has been done here. It is the micro-level PKL power plant which is feasible and viable for practical utilization in Bangladesh. In this case, the PKL extract should be filtered before using electricity production. The production method should be dynamic method instead of the static method. This micro-power plant can be used for many purposes including computer and laptop which is shown in Table-1. This PKL power can be used equally both day- and nighttime and also in the rainy season. This work will be the guideline for the betterment of mankind in near future.

Table 10 Utilization of PKL electricity with loads

Local time (hrs)	PKL module voltage with load (Volt)
00	48.42
10	48.22
15	48.29
20	48.29
25	48.29
30	48.24
35	48.17
40	48.17
45	48.17
50	48.20
55	48.25
60	48.10
65	48.12
70	47.99
75	47.80
80	47.86
85	47.79
90	47.78
95	47.77
100	47.76
105	47.75
110	47.73
115	47.71
120	47.69
125	47.69
130	47.65
135	47.65
140	47.58
145	47.55
150	47.54

Table 11 Utilization of PKL electricity with loads

Local Time (hrs)	Voltage with load (Volt)
00	49.66
05	49.65
10	49.63
15	49.60
20	49.58
25	49.53
30	49.55
35	49.53
40	49.50
45	49.47
50	49.45
55	49.43
60	49.41
65	49.40
70	49.39
75	49.38
80	49.30
85	49.27
90	49.23
95	49.19
100	49.15
105	49.11
110	49.10
115	49.03
120	49.00
125	48.97
130	48.94
135	48.91
140	48.89
145	48.88
150	48.85

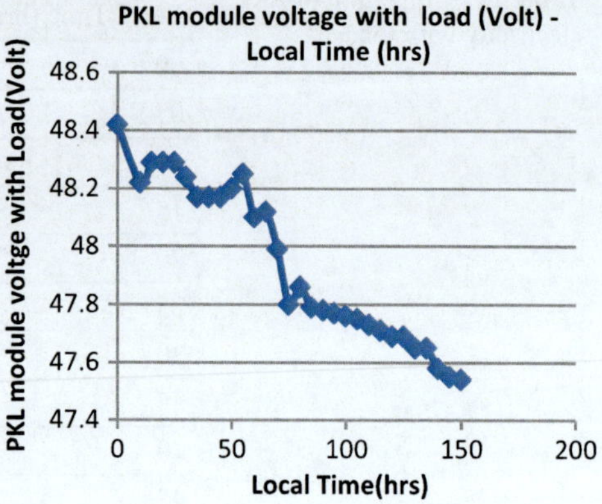

Fig. 14 PKL module voltage with load versus local time graph

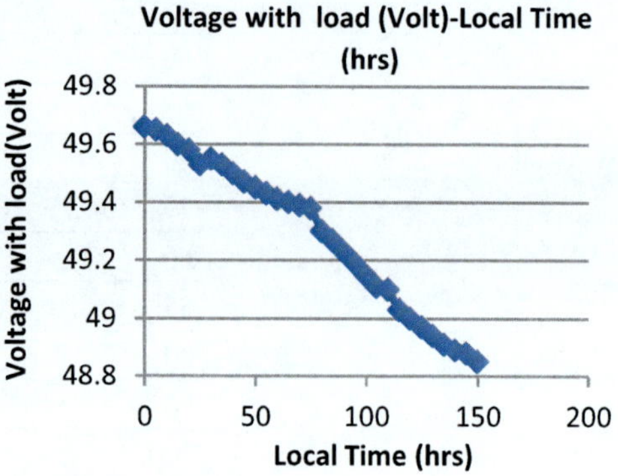

Fig. 15 PKL module voltage with load versus local time graph

Acknowledegments The authors are grateful to the PKL electricity research group named Dr. Md. Sazzad Hossain, Md. Mehedi Hasan, Dr. Jesmin Sultana, and Prof. Dr. Mesbah Uddin Ahmed for their valuable suggestions and wholehearted cooperation during research work.

References

1. Khan, K.A., Rahman Rasel, S., Ohiduzzaman, M., Homemade, P.K.L.: Electricity Generation for Use in DC Fan at Remote Areas, Accepted and is Going to be Published in Microsystem Technologies, Springer, MITE-D-19–00131 (2019)
2. Khan, K.A., Hassan, L., Obaydullah, A.K.M., Azharul Islam, S.M., Mamun, M.A., Akter, T., Hasan, M., Shamsul Alam, M., Ibrahim, M., Mizanur Rahman, M., Shahjahan, M.: Bioelectricity: A new approach to provide the electrical power from vegetative and fruits at off-grid region. J. Microsyst. Technol. Springer, manuscript number: 2018MITE-D-17–00623R2, Received: 14 August 2017/Accepted: 3 February 2018, **24**(3), Impact Factor: 1.195, ISSN: 0946–7076 (Print) 1432–1858 (Online), Springer-Verlag GmbH Germany, Part of Springer Nature. https://doi.org/10.1007/s00542-018-3808-3, 2018.
3. Hasan, M., Khan, K.A.: Dynamic Model of Bryophyllum pinnatum Leaf Fueled BPL Cell: A Possible Alternate Source of Electricity at the Off-grid Region in Bangladesh, Published in

the Microsystem Technologies (2018). Springer, manuscript number, MITE-D-18–00800R1. https://doi.org/10.1007/s00542-018-4149-y, Publisher Name: Springer Berlin Heidelberg,Print ISSN: 0946–7076,Online ISSN: 1432–1858, First Online: 28 September 2018
4. Khan, K.A., Bhuyan, M.S., Mamun, M.A., Ibrahim, M., Hassan, L., Wadud, M.A.: Organic Electricity from Zn/Cu-PKL Electrochemical Cell, Published in the Springer Nature, Series Title: Advs in Intelligent Syst., Computing, Volume Number:812, Book Title: Contemporary Advances in Innovative and Applicable Information Technology,ISBN:978–981–13–1539–8. https://doi.org/10.1007/978-981-13-1540-4, 2018
5. Khan, K.A., Hazrat Ali, M., Obaydullah, A.K.M., Wadud, M.A.: Production of candle using solar thermal technology, accepted and is going to be published in microsystem technologies. Springer, and MITE-D-19–00119 (2019)
6. Hasan, L., Hasan, M., Khan, K.A., Azharul Islam, S.M.: SEM analysis of electrodes and measurement of ionic pressure by AAS data to identify and compare the characteristics between different bio-fuel based electrochemical cell. In: The International Conference on Physics-2018. Venue-Department of Physics, University of Dhaka, Dhaka-1000,Bangladesh, Organizer-Bangladesh Physical Society(BPS), pp. 46 (2018)
7. Khan, K.A., Rahman, A., Rahman, M.S., Tahsin, A., Jubyer, K.M., Paul, S.: Performance analysis of electrical parameters of PKL electricity (An experimental analysis on discharge rates, capacity and discharge time, pulse performance and cycle life and deep discharge of PathorKuchi Leaf (PKL) electricity cell). In: Innovative Smart Grid Technologies-Asia (ISGT-Asia), 2016 IEEE, pp. 540–544. IEEE (2016)
8. Khan, M.K.A., Paul, S., Rahman, M.S., Kundu, R.K., Hasan, M.M., Moniruzzaman, M., Mamun, M.A.: A study of performance analysis of PKL electricity generation parameters: (An experimental analysis on voltage regulation, capacity and energy efficiency of pathorkuchi leaf (PKL) electricity cell). In: Power India International Conference (PIICON), 2016 IEEE 7th, pp. 1–6. IEEE (2016)
9. Khan, M.K.A., Rahman, M.S., Das, T., Ahmed, M.N., Saha, K.N., Paul, S.: Investigation on parameters performance of Zn/Cu electrodes of PKL, AVL, Tomato and lemon juice based electrochemical cells: a comparative study. In: 2015 3rd International Conference on Electrical Information and Communication Technology (EICT), pp. 1–6. IEEE (2017)
10. Tahsin, A., Jubyer, K.M., Paul, S.: Performance analysis of electrical parameters of PKL electricity (An experimental analysis on discharge rates, capacity and discharge time, pulse performance and cycle life and deep discharge of Pathor Kuchi Leaf (PKL) electricity cell). In: 2016 IEEE on Innovative Smart Grid Technologies-Asia (ISGT-Asia), pp. 540–544. IEEE (2016)
11. Khan, M.K.A., Paul, S., Rahman, M.S., Kundu, R.K., Hasan, M.M., Moniruzzaman, M., Al Mamun, M.: A study of performance analysis of PKL electricity generation parameters: (An experimental analysis on voltage regulation, capacity and energy efficiency of pathor kuchi leaf (PKL) electricity cell). In: 2016 IEEE 7th on Power India International Conference (PIICON), pp. 1–6. IEEE (2016)
12. Khan, M.L.A.: Copper oxide coating for use in linear solar fresnel reflecting concentrating collector. J. Elsevier Renew. Energy Int. J. WREN(World Renewable Energy Network), UK, RE: 12.97/859 (1998)
13. Hamid, M.R.: Characterization of a battery cell fueled by bryophyllum pinnatum sap. Int. J. Sci. Eng. Res. **4**(3), 1–4 (2013). ISSN 2229–5518
14. Hamid, M.R., Yusuf, A., Abdul Wadud, A.M., Rahaman, M.M.: Design and Performance Test of a Prototype of a 12 Volt DC Battery Fueled by Bryophyllum Pinnatum Sap and Improvement of Its Characteristics, Department of Electrical and Electronic Engineering, Ahsanullah University of Science and Technology, Dhaka, Bangladesh. Int. J. Electron. Electr. Eng. **4**(5), 398–402 (2016)
15. Khan, M.K.A., Rahman, M.S., Das, T., Ahmed, M.N., Saha, K.N., Paul, S.: Investigation on parameters performance of Zn/Cu Electrodes of PKL, AVL, tomato and lemon juice based electrochemical cells: a comparative study. In: 2017 3rd International Conference on Electrical Information and Communication Technology (EICT), pp. 1–6. IEEE, Khulna, Bangladesh, Bangladesh (2017). DOI: https://doi.org/10.1109/EICT.2017.8275150,IEEE

16. Khan, K.A., Arafat, M.E.: Development of Portable PKL (Pathor Kuchi Leaf) Lantern. Int. J. SOC. Dev. Inf. Syst. **1**(1): 15–20 (2010).
17. Khan, K.A., Bosu, R.: Performance study on PKL electricity for using DC Fan. Int. J. SOC. Dev. Inf. Syst. **1**(1): 27–30 (2010)
18. Khan, K.A., Hossain, M.I.: PKL Electricity for switching on the television and radio. Int. J. SOC. Dev. Inf. Syst. **1**(1): 31–36 (2010)
19. Paul, S., Khan, K.A., Islam, K.A., Islam, B., Reza, M.A.: Modeling of a Biomass Energy based (BPL) generating power plant and its features in comparison with other generating plants .IPCBEE **44**: @ (2012) IACSIT Press. Singapore (2012). https://doi.org/10.7763/IPCBEE.2012.V44.3
20. Khan, K.A., Paul, S., Adibullah, M., Alam, M.F., Sifat, S.M., Yousufe, M.R.: Performance analysis of BPL/PKL Electricity module. Int. J. Sci. Eng. Res. **4**(3), 1–4 (2013). 1 ISSN 2229–5518
21. Khan, K.A., Paul, S., Zobayer, A., Hossain, S.S.: A study on solar photovoltaic conversion. Int. J. Sci. Eng. Res. *4*(3), 1–5 (2013). ISSN2229–5518
22. Akter, T., Bhuiyan, M.H., Khan, K.A.: Impact of photo electrode thickness and annealing temperature on natural dye sensitized solar cell. J. Elsevier. Ms. Ref. No.: SETA-D-16–00324R2 (2017)
23. Khan, K.A.: Inventors, Electricity Generation form Pathor Kuchi Leaf (PKL), Publication date 2008/12/31, Patent number BD 1004907 (2008)
24. Khan, K.A.: Technical note Copper oxide coatings for use in a linear solar Fresnel reflecting concentrating collector. Publication date 1999/8/1, J. Renew. Energy **17**(4), 603–608 (1999). Publisher–Pergamon
25. Khan, K.A., Paul, S.: A analytical study on Electrochemistry for PKL (Pathor Kuchi Leaf) electricity generation system. In: Publication date 2013/5/21, Conference- Energytech, 2013 IEEE, pp. 1–6, IEEE (2013)
26. Ruhane, T.A., Islam, M.T., Rahaman, M.S., Bhuiyan, M.M.H., Islam, J.M.M., Newaz, M.K., Khan, K.A., Khan, M.A.: Photo current enhancement of natural dye sensitized solar cell by optimizing dye extraction and its loading period. Published in the journal of Elsevier : Optik—Int. J. Light Electron Opt. (2017)
27. Khan, K.A., Alam, M.S., Mamun, M.A., Saime, M.A., Kamal, M.M.: Studies on electrochemistry for Pathor Kuchi Leaf Power System, Ppublished in the Journal of Bangladesh. J. Agric. And Envirin. **12**(1), 37–42 (June)
28. Hasan, M., Hassan, L., Haque, S., Rahman, M., Khan, K.A.: A Study to analyze the self-discharge characteristics of bryophyllum pinnatum leaf fueled BPL Test Cell. Published in the J. IJRET **6**(12), 6–12 (2017)
29. Sultana, J., Khan, K.A., Ahmed, M.U.: Electricity Generation From Pathor Kuchi Leaf (PKL) (Bryophillum Pinnatum). J. Asiat Soc. Bangladesh Sci. **37**(4), 167–179 (2011)
30. Hasan, M., Haque, S., Khan, K.A.: An experimental study on the coulombic efficiency of bryophyllum pinnatum leaf generated BPL cell. IJARIIE **2**(1), 194–198 (2018). ISSN(O)-2395–4396
31. Hasan, M.M., Khan, M.K.A., Khan, M.N.R., Islam, M.Z.: Sustainable Electricity Generation at the Coastal Areas and the Islands of Bangladesh Using Biomass Resources. City University Journal **02**(01), 09–13 (2016)
32. Hasan, M., Khan, K.A.: Bryophyllum pinnatum leaf fueled cell: an alternate way of supplying electricity at the off-grid areas in Bangladesh. In: Proceedings of 4th International Conference on the Developments in Renewable Energy Technology [ICDRET 2016], p. 01 (2016). https://doi.org/10.1109/ICDRET.2016.7421522
33. Hasan, M., Khan, K.A., Mamun, M.A.: "An Estimation of the Extractable Electrical Energy from Bryophyllum pinnatum Leaf", American International Journal of Research in Science, Technology. Engineering & Mathematics (AIJRSTEM) **01**(19), 100–106 (2017)
34. Khan, K.A.: Electricity Generation form Pathor Kuchi Leaf (*Bryophyllum pinnatum*). Int. J. Sustain. Agril. Tech. **5**(4), 146–152 (July)

35. Hossain, M.A., Khan, M.K.M., Quayum, M.E.: 'Performance development of bio-voltaic cell from arum leaf extract electrolytes using zn/cu electrodes and investigation of their electrochemical performance. Int. J. Adv. Sci. Eng. Technol. **5**(4)(Spl. Issue 1) (2017). ISSN: 2321–9009
36. Khan, K.A., Wadud, M.A., Obaydullah, A.K.M., Mamun, M.A.: PKL (Bryophyllum Pinnatum) electricity for practical utilization. IJARIIE **4**(1), 957–966 (2018). ISSN(O)-2395–4396
37. Haque, M.M., Ullah, A.K.M.A., Khan, M.N.L., Kibria, A.K.M.F.F., Khan, K.A.: Phytosynthesis of MnO2 Nanoparticles for generating electricity. In the International Conference on Physics-2018, Venue-Department of Physics, University of Dhaka, Dhaka-1000,Bangladesh, Organizer-Bangladesh Physical Society (BPS) (2018)
38. Hasan, M., Khan, K.A.: Identification of BPL cell parameters to optimize the output performance for the off-grid electricity production. In the International Conference on Physics-2018, Venue-Department of Physics, University of Dhaka, Dhaka-1000,Bangladesh, Organizer-Bangladesh Physical Society(BPS), pp. 60, (2018)
39. Khan, K.A., Bhuyan, M.S., Mamun, M.A., Ibrahim, M., Hassan, L., Wadud, M.A.: Organic electricity from Zn/Cu-PKL electrochemical cell. In: Souvenir of First International Conference of Contemporary Advances in Innovative & Information Technology(ICCAIAIT) (2018). Organized by KEI, In collaboration with Computer Society of India(CSI), Division-IV(Communication). pp. 75–90. The proceedings consented to be published in AISC Series of Springer (2018)
40. Khan, M.K.A., Obaydullah, A.K.M., Wadud, M.A., Hossain, M.A.: Bi-Product from Bioelectricity. IJARIIE **4**(2), 3136–3142 (2018). ISSN(O)-2395–4396
41. Khan, M.K.A., Obaydullah, A.K.M.: Construction and commercial Use of PKL Cell. IJARIIE **4**(2), 3563–3570 (2018). ISSN(O)-2395–4396
42. Khan, M.K.A.: Studies on Electricity Generation from Stone Chips Plant (Bryophyllum pinnatum). International J. Eng. Tech **5**(4), 393–397 (Dece)
43. Khan, K.A., Hossain, M.A., Obaydullah, A.K.M., Wadud, M.A.: PKL electrochemical cell and the Peukert's Law. IJARIIE **4**(2), 4219 – 4227 (2018). ISSN(O)-2395–4396
44. Khan, K.A., Wadud, M.A., Hossain, M.A., Obaydullah, A.K.M.: Electrical Performance of PKL (Pathor Kuchi Leaf) Power. IJARIIE **4**(2), 3470–3478 (2018). ISSN(O)-2395–4396
45. Khan, K.A., Ali, M.H., Mamun, M.A., Haque, M.M., Atique Ullah, A.K.M., Islam Khan, M.N., Hassan, L., Obaydullah, A.K.M., Wadud, M.A.: Bioelectrical characteristics of Zn/Cu-PKL Cell and Production of Nanoparticles (NPs) for Practical Utilization. In: 5th International conference on 'Microelectronics, Circuits and Systems', Micro 2018, 2018,Venue: Bhubaneswar, Odisha, India, Organizer: Applied Computer Technology, Kolkata, West Bengal, India (2018). pp. 59–66. www.actsoft.org. ISBN: 81–85824–46–1, In Association with: International Association of Science, Technology and Management, 2018
46. Hassan, M.M., Arif, M., Khan, K.A.: Modification of Germination and growth patterns of Basella alba seed by low pressure plasma. J. Modern Phys. Paper ID: 7503531 (2018)
47. Khan, K.A., Maniruzzaman Manir, S.M., Islam, M.S., Jahan, S., Hassan, L., Ali, M.H.: Studies on nonconventional energy sources for electricity generation. Int. J. Adv. Res. Innovat. Ideas Edu **4**(4), 229–244 (2018)
48. K A Khan, Mahmudul Hasan, Mohammad Ashraful Islam, Mohammad Abdul Alim, Ummay Asma, Lovelu Hassan, and M Hazrat Ali. "A Study on Conventional Energy Sources for Power Production" Internation Journal Of Advance Research And Innovative Ideas In Education, Volume 4 Issue 4 2018 Page 214–228
49. Khan, M.K.A., Rahman, M.S., Das, T., Ahmed, M.N., Saha, K.N., Paul, S.: Investigation on parameters performance of Zn/Cu electrodes of PKL, AVL, Tomato and Lemon juice based electrochemical cells: A comparative study, Publication Year: 2017, Page(s):1–6, Published In: 2017 3rd International Conference on Electrical Information and Communication Technology (EICT), Date of Conference: 7–9 Dec. 2017, Date Added to IEEE Xplore: 01 February 2018, ISBN Information: INSPEC AccessionNumber: 17542905,DOI: https://doi.org/10.1109/EICT.2017.8275150,Publisher: IEEE,Conference Location: Khulna, Bangladesh

50. Khan, K.A., Alam, M.M.: Performance of PKL (Pathor Kuchi Leaf) Electricity and its Uses in Bangladesh. Int. J. SOC. Dev. Inf. Syst. **1**(1), 15–20 (Janu)
51. Khan, K.A., Bakshi, M.H., Mahmud, A.A.: Bryophyllum Pinnatum leaf (BPL) is an eternal source of renewable electrical energy for future world. Am. J. Phys. Chem. **3**(5), 77–83 (2014). published, online, November10, 2014 http://www.sciencepublishinggroup.com/j/ajpc, https://doi.org/10.11648/j.ajpc.20140305.15,ISSN:2327-2430 (Print); ISSN: 2327–2449 (Online)
52. Khan, M.K.A.: An experimental observation of a PKL electrochemical cell from the power production view point. In: Presented as an Invited speaker and Abstract Published in the Conference on Weather Forecasting and Advances in Physics (2018). Department of Physics, Khulna University of Engineering and Technology (KUET), Khulna, Bangladesh.
53. Guha, B., Islam, F., Khan, K.A.: Studies on redox equilibrium and electrode potentials. IJARIIE **4**(4), 1092–1102 (2018). ISSN(O)-2395–4396
54. Islam, F., Guha, B., Khan, K.A.: Studies on pH of the PKL extract during electricity generation for day and night time collected Pathor Kuchi Leaf. IJARIIE **4**(4), 1102–1113 (2018). ISSN(O)-2395–4396
55. Khan, K.A., Rahman, M.L., Islam, M.S., Latif, M.A., Khan, M.A.H., Saime, M.A., Ali, M.H.: Renewable energy scenario in Bangladesh. J. IJARII **4**(5), 270–279 (2018). ISSN(O)-2395–4396
56. Khan, K.A., Rasel, S.R.: Prospects of renewable energy with respect to energy reserve in Bangladesh. J. IJARII **4**(5), 280–289 (2018). ISSN(O)-2395–4396
57. Khan, K.A., Hossain, M.S., Kamal, M.M., Rahman, M.A., Miah, I.: Pathor Kuchi Leaf : importance in power production. IJARIIE **4**(5) (2018). ISSN(O)-2395–4396
58. Khan, K.A., Hazrat Ali, M., Mamun, M.A., Ibrahim, M., Obaidullah, A.K.M., Afzol Hossain, M., Shahjahan, M.: PKL electricity in mobile technology at the off-grid region. Published in the Proceedings of CCSN-2018, 2018 at Kolkata, India.(2018)
59. Khan, K.A., Hossain, A.: Off-grid 1 KW PKL Power Technology: design, fabrication, installation and operation. In: Published in the proceedings of CCSN-2018, 2018 at Kolkata, India (2018)
60. Khan, K.A., Mamun, M.A., Ibrahim, M., Hasan, M., Ohiduzzaman, M., Obaidullah, A.K.M., Wadud, M.A., Shajahan, M.: PKL electrochemical cell for off-grid Areas: Physics, Chemistry and Technology, Published in the proceedings of CCSN-2018, 2018 at Kolkata, India (2018)
61. Khan, K.A., Rahman Rasel, S.: Studies on Wave and Tidal Power Extraction Devices. Int. J. Ad. Res. Innovat. Ideas Edu. 4(60), 61–70 (2018)
62. Khan, K.A., Ahmed, S.M., Akhter, M., Rafiqul Alam, M., Hossen, M.: Wave and tidal power generation. Int. J. Adv. Res. Innovat. Ideas Edu. 4(6), 71–82 (2018)
63. Khan, K.A., Atiqur Rahman, M., Nazrul Islam, M., Akter, M., Shahidul Islam, M.: Wave Climate study for ocean power extraction. Int. J. Adv. Res. Innovat. Ideas Edu. 4(6), 83–93 (2018)
64. Khan, K.A., Sujan Miah, M., Iman Ali, M., Sharma, S.K., Quader, A.: Studies on Wave and Tidal Power Converters for Power Production. Int. J. Adv. Res. Innovat. Ideas Edu. 4(6), 94–105 (2018)
65. Khan, K.A., Yesmin, F.: PKL Electricity- A Step forward in Clean Energy. Int. J. Adv. Res. Innovat. Ideas Edu. 5(1), 316–325 (2019)
66. Khan, K.A., Hazrat Ali, M., Obaydullah, A.K.M., Wadud, M.A.: Candle production using solar thermal systems. In: 1st International Conference on 'Energy Systems, Drives and Automations', ESDA2018, pp. 55–66
67. Khan, K.A., Yesmin, F.: Cultivation of electricity from living PKL Tree's leaf. Int. J. Adv. Res. Innovat. Ideas Edu. **5**(1), 462–472 (2019)
68. Khan, K.A., Rahman Rasel, S., Ohiduzzaman, M.: Homemade PKL Electricity Generation for Use in DC Fan at Remote Areas. In: 1st International Conference on 'Energy Systems, Drives and Automations', ESDA2018, pp. 90–99
69. Khan,K.A., Yesmin, F.: Solar water pump for vegetable field under the climatic condition in Bangladesh. Int. J. Adv. Res. Innovat. Ideas Edu. **5**(1), 631–641 (2019)

70. Khan, K.A., Rahman Rasel, S.: Solar photovoltaic electricity for irrigation under Bangladeshi climate. Int. J. Adv. Res. Innovat. Ideas Edu. **5**(2), 28–36 (2019)
71. Khan, K.A., Rahman Rasel, S.: The present scenario of nanoparticles in the world. Int. J. Adv. Res. Innovat. Ideas Edu. **5**(2), 462–471 (2019)
72. Khan, K.A., Yesmin, F., Abdul Wadud, M., Obaydullah, A.K.M.: Performance of PKL Electricity for Use in Television. In: International Conference on Recent Trends in Electronics and Computer Scienc-2019, p. 69. Venue: NIT Silchar, Assam, India, Conference date: 18th and 19th of March, 2019. Organizer: Department of Electronics and Engineering, NIT Silchar, Assam, India (2019)
73. Mamun, M.A., Ibrahim, M., Shahjahan, M., Khan, K.A.: Electrochemistry of the PKL Electricity. In: International Conference on Recent Trends in Electronics & Computer Scienc-2019, Venue: NIT Silchar, Assam, India, Conference date: 18th and 19th of March, 2019. Organizer: Department of Electronics and Engineering, p. 71. NIT Silchar, Assam, India (2019)
74. Khan, K.A., Anowar Hossain, M., Alamgir kabir, M., Akhlaqur Rahman, M., Lipe, P.: A Study on Performance of Ideal and Non-ideal Solar Cells under the Climatic Situation of Bangladesh. Int. J. Adv. Res. Innovat. Ideas Edu. **5**(2), 975–984 (2019)
75. Ohiduzzaman, M., Khan, K.A., Yesmin, F., Salek, M.A.: Studies on Fabrication and performance of solar modules for practical utilization in Bangladeshi Climate. IJARIIE **5**(2), 2626–2637 (2019)
76. Khan, K.A., Rahman Rasel, S.: A study on electronic and ionic conductor for a PKL electrochemical cell. IJARIIE **5**(2):3100–3110 (2019)
77. Ohiduzzaman, M., Khatun, R., Reza, S., Khan, K.A., Akter, S., Uddin, M.F., Ahasan, M.M.: Study of exposure rates from various nuclear medicine scan at INMAS. Dhaka. IJARIIE **5**(3), 208–218 (2019)
78. Khan, K.A., Rasel, S.R.: Development of a new theory for PKL electricity using Zn/Cu electrodes: per pair per volt. IJARIIE **5**(3), 1243–1253 (2019)
79. Khan, K.A., Abu Salek, M.: A Study on Research. Development and Demonstration of Renewable Energy Technologies, IJARIIE **5**(4), 113–125 (2019)
80. Khan, K.A., Uddin, M.N., Nazrul Islam, Md., Mondol, N., Ferdous, Md.: A study on some other likely renewable sources for developing countries. IJARIIE **5**(4), 126–134 (2019)
81. Khan, K.A., Zian Reza, S.M.: The Situation of renewable energy policy and planning in developing countries. IJARIIE **5**(4), 557–565 (2019)
82. Khan, K.A., Rasel, S.R., Reza, S.M.Z., Yesmin, F.: Electricity from Living PKL Tree, Published in the Open Access book, "Energy Efficiency and Sustainability in Outdoor Lighting A Bet for the Future" edited by Prof. Manuel J, Hermoso-Orzáez, London, UK (2019)
83. Khan, K.A., Salek, M.A.: Solar Photovoltaic (SPV) Conversion: A Brief Study. IJARIIE **5**(5), 187–204 (2019)
84. Hassan, L., Khan, K.A.: A Study on Harvesting of PKL Electricity. Microsyst. Technol. (2019). https://doi.org/10.1007/s00542-019-04625-7
85. Hasan, M., Khan, K.A.: Experimental characterization and identification of cell parametersin a BPLelectrochemicaldevice. SN Appl.Sci. **1**:1008 (2019).https://doi.org/10.1007/s42452-019-1045-8
86. Khan, K.A., Nusrat Zerin, S.M., Chy, N., Nurul Islam, M., Bhattacharjee, R.: A study on voltage harvesting from PKL living plant. IJARIIE **5**(5), 407–415 (2019)
87. Khan, K.A., Mamun, M.A., Ibrahim, M., Hasan, M., Ohiduzzaman, M., Obaydullah, A.K.M., Wadud, M.A., Shajahan, M.: PKL electrochemical cell: physics and chemistry. SN Appl. Sci. **1**:1335 (2019).https://doi.org/10.1007/s42452-019-1363-x
88. Rab, M.N.F., Khan, K.A., Rasel, S.R., M Ohiduzzaman, M., Yesmin, F., Hassan, L., Abu Salek, M., Zian Reza, S.M., Hazrat Ali, M.: Voltage cultivation from fresh leaves of air plant, climbing spinach, mint, spinach and Indian pennywort for practical utilization. In: 8th International Conference on CCSN2019, vol 1. Institute of Aeronautical Engineering, Hyderabad, India (2019).
89. Khan, K.A., Rahman, M.S., Akter, A., Hoque, M.S., Khan, M.J., Mirja, E., Howlader, M.N., Solaiman, M.: A study on the effect of embedded surface area of the electrodes for voltage collection from living PKL tree. IJARIIE **5**(6) (2019). ISSN(O)-2395–4396

Analysis of Lakes Over the Period of Time Through Image Processing

Sattam Ghosal, Abhishek Karmakar, Pushkar Sahay, and Uma Das

Abstract The Great Salt Lake, located in the state of Utah, is the largest salt water lake in the West and the eighth largest terminal lake in the world. The lake is a home for millions of native birds, shorebirds, and waterfowl, including the largest staging population of Wilson's phalarope in the world. Since 1847, there has been a constant decline in its area. Lake Powell is a lake near the Colorado River in the United States and serves as a major vacation destination for approximate 2 million people annually. However, because of successive droughts and insatiable requirement of water for human and agriculture, Lake Powell has fallen way below in terms of water, depth, and surface area. This study focused on the temporal changes of these two lakes using the multi-temporal Landsat images using Prewitt edge detection and the comparison of the result with other edge detection techniques. The results thus obtained reflect the constant decline in Lake Powell and the Great Salt Lake in terms of surface area in the period 1984–2018 and 1987–2018, respectively.

Keywords Delineation · Landsat images · Prewitt operator · Robert operator

1 Introduction

Lakes are generally located in the rift zones, mountainous areas, and areas with continuous melting of snow near glaciers. Some lakes are found along the courses

S. Ghosal (✉) · A. Karmakar · P. Sahay · U. Das
Indian Institute of Information Technology Kalyani, Kalyani 741235, West Bengal, India
e-mail: sattam@iiitkalyani.ac.in

A. Karmakar
e-mail: abhishekkar@iiitkalyani.ac.in

P. Sahay
e-mail: pushkar_bt17@iiitkalyani.ac.in

U. Das
e-mail: uma@iiitkalyani.ac.in

J. K. Mandal et al. (eds.), *Computational Intelligence and Machine Learning*, Advances in Intelligent Systems and Computing 1276,
https://doi.org/10.1007/978-981-15-8610-1_18

of mature rivers. There are many lakes, in different parts of the globe, which were formed because of improper drainage systems. None of the lakes are permanent, as they will slowly get filled up with sediments or tip over the basin containing them. Many lakes are artificial in nature and have been constructed for serving the industrial or agricultural needs, or for hydro-electric power generation and domestic water supply, or for exquisite, recreational purposes, or some other activities.

Over the years, surface water bodies like lakes, ponds, reservoirs, etc. have been treated as a community resource. They were being nurtured, protected, conserved and managed by the major percentage of the local community without any code of conduct or rule. In turn, these water bodies have been catering the local human and livestock populations. In the modern times, after the introduction of water supply for the public and ground water usage through the hand pumps and wells, a dramatic shift in the attitude of the people towards these water bodies has been witnessed. Both the governments as well as the locals have begun ignoring this asset in the fad and fantasy of the introduced public water supply. They have just started neglecting and have really stopped being responsible for these community resources. Other than this, blooming urban and industrial development has changed the position of these water bodies from a public resource to just a dumping ground for construction debris, garbage, sewage, religious offering etc. These water bodies had fallen a prey to administrative and social atrocities. All this has put the existence of these water bodies on stake and has led to severe deterioration of their water quality. In the recent years, urgent need to restore these community resources has been realized by various countries. This is because mushrooming population and development activities have put immense strain on the public water supply and ground water extraction. This has widened demand–supply gap and has led to excessive depletion of ground water.

The present study deals with the narrowing of one of the major surface water bodies i.e. Lakes. Examining the changes in their overall structure and area done through the graphical analysis of the Landsat images with the help of our proposed framework.

2 Materials and Methods

The basic methodology adopted for carrying out the present study is primarily by analyzing and interpreting the primary data along with the review on the available literature and media reference on the issue. Some of the finest works have been done for vegetation like [1] and [2] and for river extraction [3] but when it comes to using such algorithms for smaller water bodies like lakes and ponds, the noise overlapping is way too much. Primary data for the present study are the Landsat images which were gathered from https://www.earthexplorer.usgs.gov [4, 5] and the literature which was gathered from various sources. The segmentation based on threshold detection from hue histograms [6] and the NDVI method as discussed in [7] was very beneficial in finding out the vegetation or the cropped fields in a satellite image but when same technique is used for water bodies, the results aren't very promising. The various components of the methodology for the present study are as follows.

3 Great Salt Lake

The Great Salt Lake, located in the state of Utah, is the largest salt water lake in the West and the 8th largest terminal lake in the world [8]. In an average year the lake covers an area of around 1,700 square miles (4,400 km2). In 1988, the surface area was at its peak of 3,300 square miles (8,500 km2) but there has been a constant decline ever since. It is the largest lake in the United States, in terms of surface area coverage, which is not a part of the Great Lakes region. Great Salt Lake is salty because it does not have an outlet. Tributary rivers are constantly bringing in small amounts of salt dissolved in their fresh water flow. Once in the Great Salt Lake much of the water evaporates leaving the salt behind. These large changes in water levels is the result of years of human activities such as diverting the river water, which was supposed to fill the lake, for agriculture and industry. It is approximated that about 40% of river water is usually diverted from the lake. These activities, in addition to the ongoing drought in the Western hemisphere, have drained humongous volume of water from the historic lake. As a result, evaporation from the lake's surface is significantly faster than the river inflow. The wildlife habitat is also disappearing and the health of the two million people living in the surrounding area is threatened from airborne dust coming from the dried lake-bed (Figs. 1 and 2).

Fig. 1 Salt lake 1999

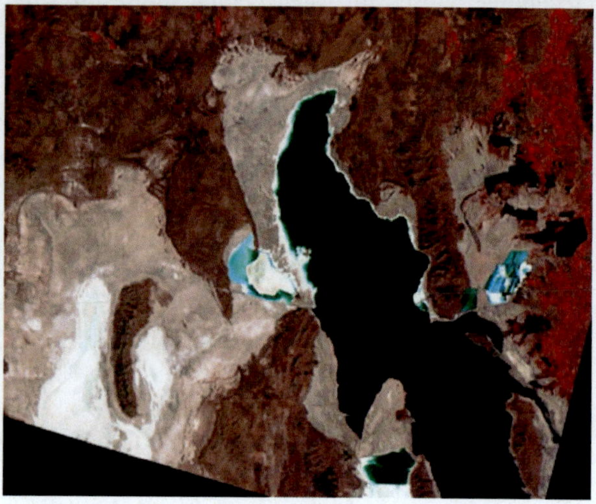

Fig. 2 Salt lake 2018

4 Lake Powell

Lake Powell situated near the Colorado river, is a lake bestriding the boundaries between Utah and Arizona in the United States [9]. Colorado River flows in from the east around Mille Crag Bend and is engulfed by the lake. At the west end of Narrow Canyon, the Dirty Devil River joins the lake from the north. It is a major vacation spot that around two million people visit every year. However, due to high water consumption for agricultural purpose and humans, and because of continuous droughts in the area, Lake Powell has fallen way below in terms of water, depth and surface area. The downfall was visible in the canyons on either side feeding the reservoir. By 2002, the water levels had dropped so much that the canyon walls, which were way too exposed, created a light- coloured outline all around the lake. Falling water levels and dry conditions are seen easily in the images of the 2000s. The side branches of the lake have all receded in comparison to the previous year's extents. In 2017, the rainfall and snowfall were as usual low, which resulted in long-term drought in the region (Figs. 3 and 4).

5 Delineation of Images

To convert an RBG image to grayscale, there are various methods to create a single value for each pixel to represents its brightness from the RBG values. One method is to take average contribution from the three channels. However, the thus obtained brightness is often influenced by the green component. To avoid that the images were delineated into RGB formats in order to analyze them individually.

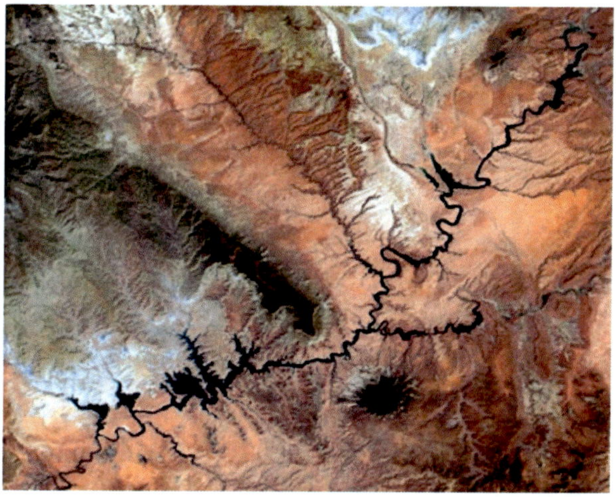

Fig. 3 Lake powell in 2018

Fig. 4 Lake powell in 1984

6 Prewitt Method

Sometimes it happens that the lakes or other water bodies can't be extracted with great accuracy whereas the coastal boundaries segmentation as discussed in the paper [10] can be achieved using spatial attraction algorithm. This might be due to the noise present in the image or the blurriness in the image itself. That's why various edge detection algorithms were used to differentiate the edge of the lakes and the other uneven parts of the images but the Prewiit edge detection method gave the best output. The above mentioned operator is a discrete differentiation operator which computes the approximated values of the image gradient and its features. It is somewhat similar to Sobel operator. It has two 3×3 kernels. Mathematically, it uses the convolution of the original image with two 3×3 kernels to get the approximate values of the derivate, namely for the horizontal and the vertical changes. The 'A' being the original image, then Gx and Gy are considered to be the two images where the vertical and horizontal approximate derivatives are stored at each point. Upon convolving the resulting gradient is given by

$$G = \sqrt{(Gx * Gx) + (Gy * Gy)}$$

And the direction of the gradient is given by

$$\theta = a \tan 2(Gx, Gy)$$

7 Difference in the Images

Lastly two differences were taken for comparing the changes in the surface area.
of the Lakes which happened over time. The differences are as follows:

- Difference in the original images (Lake Powell in 1984 and Lake Powell in 2018) to see the changes in the overall image.
- Difference in the images obtained after edge detection (rivers abstracted).

The average color of the zones and the brightness were used to compare two pictures for likeness. The basic approach for the same was as follows:

- Check dimensions. If different, then images are not the same.
- Check formats. If same, then perform precise comparison, pixel by pixel.
- If different formats do this: Compare Brightness as half the weight and compare color/hue as the other half. Calculate the difference in values and depending on'tolerance' value they are the same or they are not.

8 Results

8.1 Great Salt Lake

After calculating the differences in the original images and the images obtained after Prewitt detection were analyzed, the changes in the surface area was calculated over the years. The results were as follows:

- Change in the surface area from 1987 to 1999: –18.06%
- Change in the surface area from 1999 to 2011: –13.34%
- Change in the surface area from 2011 to 2016: –20.74%
- Change in the surface area from 2016 to 2018: –10.05%

The overall change observed from 1987 to 2018 was –53.07% (Figs. 5 and 6).

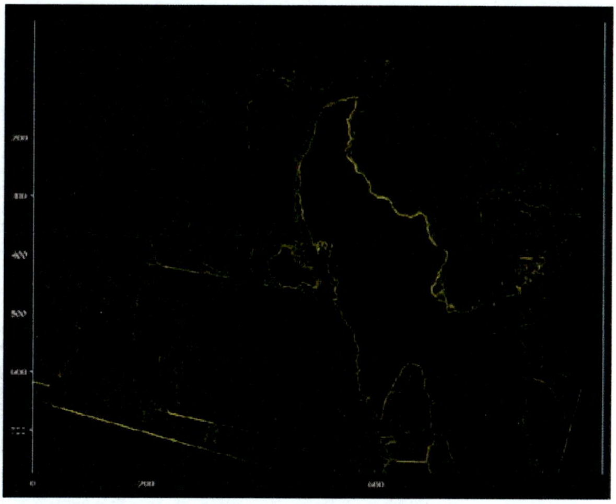

Fig. 5 Differences in the images after prewitt detection

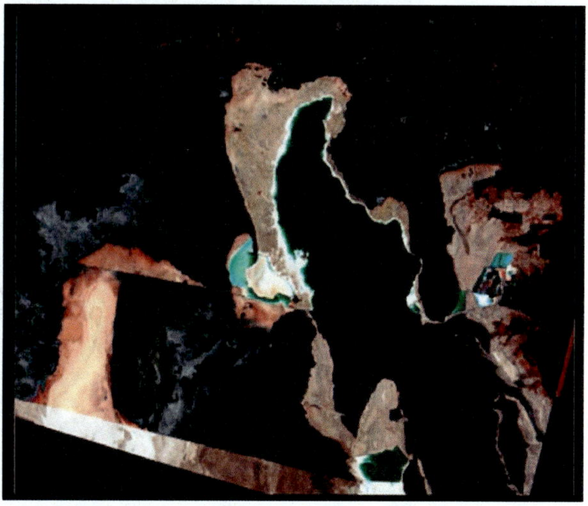

Fig. 6 Differences in the original images of 1987 and 2018

9 Lake Powell

After calculating the differences in the same way as described above, the results were as follows:

- Change in the surface area from 1984 to 1993: –12.47%
- Change in the surface area from 1993 to 1998: –26.36%
- Change in the surface area from 1998 to 2018: –35.65%

The overall change observed from 1984 to 2018 was –54.76% (Figs. 7 and 8).

According to an article by Denver Post [11] and Science Mag [12], the actual water level loss of Lake Powell was 52% and that of Great Salt Lake was 50%

Fig. 7 Differences in the images after prewitt detection

Fig. 8 Differences in the original images of 1984 and 2018

whereas our study showed it to be 54.76% and 53.07% respectively. The accuracy of our proposed framework was 94.69%. and 93.86% (Figs. 9 and 10).

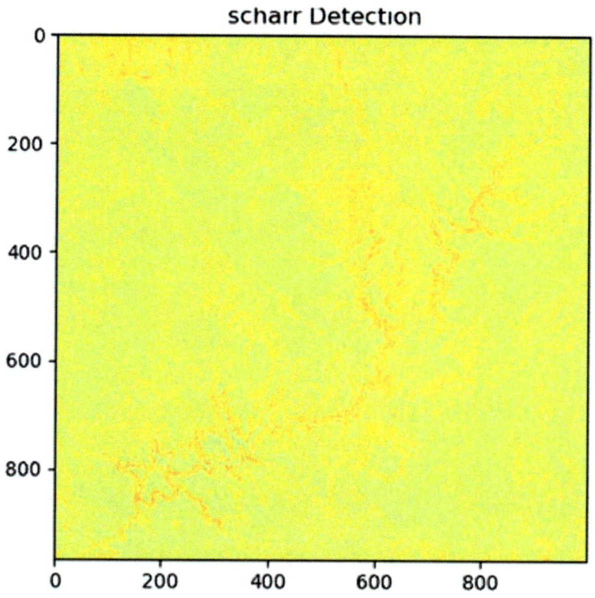

Fig. 9 Scharr detection

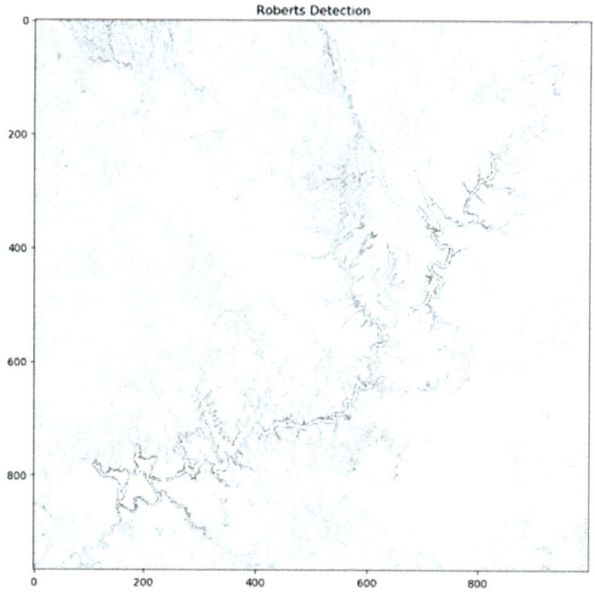

Fig. 10 Roberts detection

The decision to choose a particular edge detection technique played a vital role here. Comparisons of various edge detection techniques based on their outputs can be observed here (Figs. 11 and 12).

Fig. 11 Prewitt detection

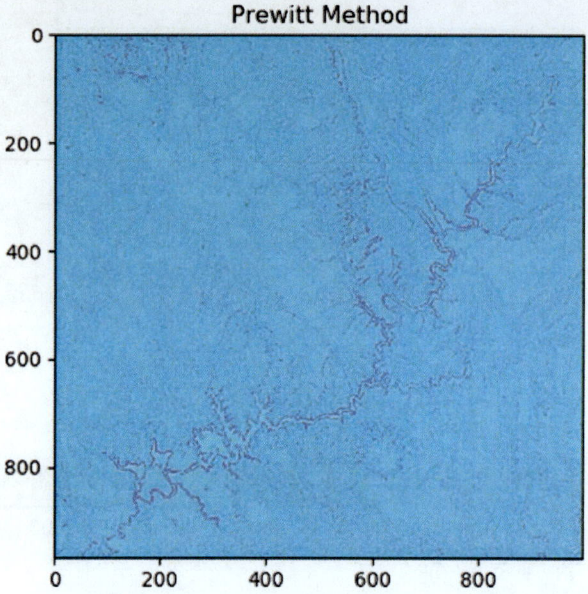

Fig. 12 Sobel detection

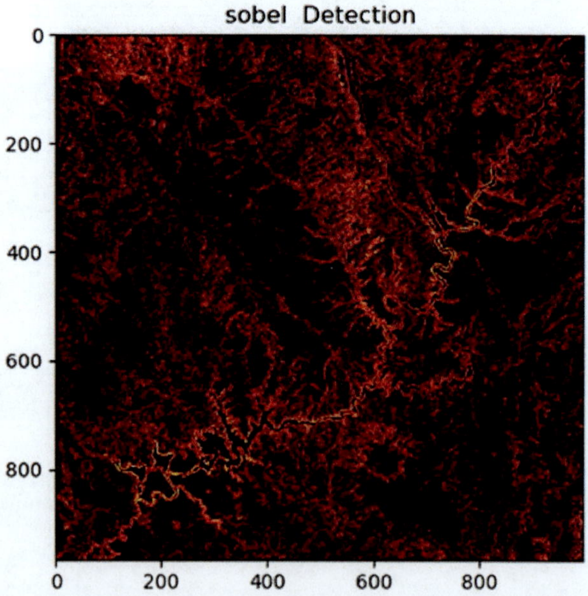

10 Summary and Discussion

Extracting water bodies from satellite images is a challenging task and has a few problems due to (1) merged water pixels, (2) scene dependent threshold levels, (3) background noise. Landsat imaging and water quality monitoring are of vital importance as it gives particular data about the quality and nature of the water bodies. In the current work we are thus able to detect the changes in surface water from images corresponding to different epochs.

As shown in the study, Lake Powell and Great Salt Lake lost more than half of this surface area in the period 1984–2018 and 1987–2018 respectively. If it continues in the same way, it is very likely that the lakes will lose all their surface area in the near future. This is very critical because the lakes provide many benefits for the society and the people living in their surroundings. Therefore, immediate and befitting measures should be taken by authorities to mitigate further decline of these lakes surface area and to restore the lakes to their original conditions. It is very clear that the construction of dams on the rivers flowing to the lakes, tremendous ground water usage, divergence of water sources to agricultural, industrial and domestic uses, and subsequent drought have all reduced the surface area of these lakes. Further, changes in watershed due to changes in rainfall and agricultural land use should also be investigated over the period of time.

References

1. Lobell, D.: Systems and methods for satellite image processing to estimate crop yield. US. Patent No. 9,953,241 (2018)
2. Todoroff, P. et al.: Automatic satellite image processing chain for near real-time sugarcane harvest monitoring. ISSCT (2018)
3. Sghaier, M.O. et al.: Combination of texture and shape analysis for a rapid rivers extraction from high resolution SAR images. In: 2016 IEEE International Geoscience and Remote Sensing Symposium (IGARSS). IEEE (2016)
4. Earthshots.usgs.gov.: Lake Powell, Utah and Arizona, USA | Earthshots: Satellite Images of Environmental Change (2019). [online] Available at: https://earthshots.usgs.gov/earthshots/node/79 .
5. Earthshots.usgs.gov.: Great Salt Lake, Utah, USA | Earthshots: Satellite Images of Environmental Change (2019). [online] Available at: https://earthshots.usgs.gov/earthshots/Great-Salt-Lake .
6. Hassanein, M., Lari, Z., El-Sheimy, N.: A new vegetation segmentation approach for cropped fields based on threshold detection from hue histograms. Sensors **18**(4), 1253 (2018)
7. Ju, J., Masek, J.G.: The vegetation greenness trend in Canada and US Alaska from 1984–2012 Landsat data. Remote Sens. Environ. **176**, 1–16 (2016)
8. Wikipedia: Great Salt Lake. https://en.wikipedia.org/wiki/Great_Salt_Lake
9. Wikipedia: Lake Powell. https://en.wikipedia.org/wiki/Lake_Powell
10. Frazier, P.S., Page, K.J.: Water body detection and delineation with Landsat TM data. Photogramm. Eng. Remote Sens. **66**, 1461–1468 (2000)

11. Denverpost.com.: Water levels drop at Lake Mead, Lake Powell amid drought (2019). [online] Available at: https://www.denverpost.com/2018/09/03/lake-mead-lake-powell-drought-colorado-river/.
12. Science|AAAS.: Utah's Great Salt Lake has lost half its water, thanks to thirsty humans (2019). [online] Available at: https://www.sciencemag.org/news/2017/11/utah-s-great-salt-lake-has-lost-half-its-water-thanks-thirsty-humans .

Video Watermarking for Persistent and Robust Tracking of Entertainment Content (PARTEC)

Deepayan Bhowmik, Charith Abhayaratne, and Stuart Green

Abstract The exploitation of film and video content on physical media, broadcast and Internet involves working with many large media files. The move to file-based workflows necessitates the copying and transfer of digital assets amongst many parties, but the detachment of assets and their metadata leads to issues of reliability, quality and security. This paper proposes a novel watermarking-based approach to deliver a unique solution to enable digital media assets to be maintained with their metadata persistently and robustly. Watermarking-based solution for entertainment content manifests new challenges, including maintaining high quality of the media content, robustness to compression and file format changes and synchronisation against scene editing. The proposed work addresses these challenges and demonstrates interoperability with an existing industrial software framework for media asset management (MAM) systems.

1 Introduction

Digital watermarking received significant attention in the recent past for various Multimedia-related applications such as copyright protection, image quality monitoring or media integrity verification. For example, Cox et al. [7] identified *broadcast*

D. Bhowmik (✉)
Division of Computing Science and Mathematics,
University of Stirling, Stirling, UK
e-mail: deepayan.bhowmik@stir.ac.uk

C. Abhayaratne
Department of Electronic and Electrical Engineering, University of Sheffield,
Sheffield, UK
e-mail: c.abhayaratne@sheffield.ac.uk

S. Green
ZOO Digital Group plc, Sheffield, UK
e-mail: stuart.green@zoodigital.com

J. K. Mandal et al. (eds.), *Computational Intelligence and Machine Learning*, Advances in Intelligent Systems and Computing 1276,
https://doi.org/10.1007/978-981-15-8610-1_19

monitoring, *owner identification*, *proof of ownership*, *authentication*, *transactional watermarking*, *copy control* and *covert communication* as potential watermarking applications of which some are adopted in the industry[1] with additional applications such as *audience measurement* or *improved auditing*. Amongst other applications (1) image quality measurement was proposed in [10] by measuring the degradation of the extracted watermarking; (2) [12] proposed a watermarking method for improved management of medical records through embedding patient data such as identity, serial number or region of interest to ensure the image association with the correct patient; or (3) a real-time system for video-on-demand services where frame images are watermarked, unique to the user and aims to deter piracy [15]. On contrary, this paper considers a new application of watermarking in tracking entertainment content from production to post-processing to distribution. In designing our new algorithm, it is important to revisit the literature that looked at various characteristics of watermarking algorithms including algorithms that are either imperceptible [1], robust to intentional [8] and unintentional (e.g., compression [11], filtering or geometric [16]) attacks or fragile [6] or secure [4].

This work proposes a new video watermarking application for Persistent and Robust Tracking of Entertainment Content (PARTEC). PARTEC is concerned with improving the assets and their metadata management in file-based workflows required for reliably copying and transferring digital assets amongst many parties. The exploitation of film and video content on physical media, broadcast and the Internet involves working with many large media files. Because of the size of media files, they are inevitably copied multiple times during post-production so that operators and different physical locations can work efficiently by having a local copy. This means that it is not practical to record the assets in one central location from which multiple operators have common access. A consequence of this is that multiple copies exist and therefore there is difficulty in maintaining the integrity of those copies. For example, if a change is made to the *master* copy, then there is currently no simple way for that change to be propagated to all copies, i.e., version control.

Detachment of assets and their metadata leads to issues of reliability, quality and security in media post-processing industry. PARTEC delivers a new, unique watermarking-based solution to enable digital media assets to be maintained and tracked with their metadata persistently and robustly by embedding unique identifiers as a watermark. Additionally, the proposed solution also enables personalisation of each copy of an asset file and so affords a level of security to prevent unauthorised access to protected data. However, watermarking entertainment content manifests new challenges, such as (a) limited/no distortion (due to embedding) is permissible for high-resolution and high-quality studio contents; (b) the solution must be compatible with existing industrial file formats; (c) robust to reasonable compression and format changes; and (d) robust to scene editing, i.e., inclusion or exclusions of frames within existing content, joining multiple clips to create a new edit or producing multiple clips from a single source. Our work proposes a solution that addresses the above-mentioned challenges and demonstrates interoperability with an existing

[1] www.digitalwatermarkingalliance.org.

industrial software framework for media asset management (MAM) systems. Main contributions of our work are

- A unique watermarking-based system for persistent and robust tracking of entertainment content,
- New watermarking algorithms suitable for frame domain, MPEG-2 and H.264 compression domain embedding with high imperceptibility and
- Techniques for watermark identifier extraction robust to scene editing, media file format changes and compression.

2 Motivation and Requirements Analysis

PARTEC considers wider industry requirements from the users including (a) media companies producing film and video content; (b) content post-processing companies providing professional services to entertainment markets; and (c) developers of MAM systems for the entertainment industry. Application scenarios include a media company wishes to offer paid for video-based educational resources which are available to consumers across the world through its video streaming website. Updatable metadata is a critical component of these materials, is time-sensitive and easily detached from the media when teachers download and create copies, thereby diminishing value. Our solution ensures the persistent association of media with the associated metadata throughout commercialisation of the media content. The approach delivers the very latest metadata on each occasion when a media asset is accessed and enables authentication of file based assets by validating their contents, version, the associated access permissions, rights clearances and other metadata including any revisions to an asset's metadata.

PARTEC represents a new application of watermarking techniques for the purpose of asset tracking, authentication and security. To the best knowledge of the authors, there exists no solution that addresses this scope or that applies imperceptible watermarking techniques for the purposes envisaged, having consulted with significant content producers and assessing the product specifications of MAM systems. As a significant shift towards file-based workflows in the entertainment industry mandates for greater levels of tracking and security, our approach is timely and solves an existing industrial problem. An overall system diagram is depicted in Fig. 1 and this paper describes the watermarking-based solution (coloured boxes) which is the central theme of PARTEC.

2.1 Requirements Analysis

As the part of the system design, we analyse the requirements with a greater understanding of the approaches currently taken in the industry for media asset manage-

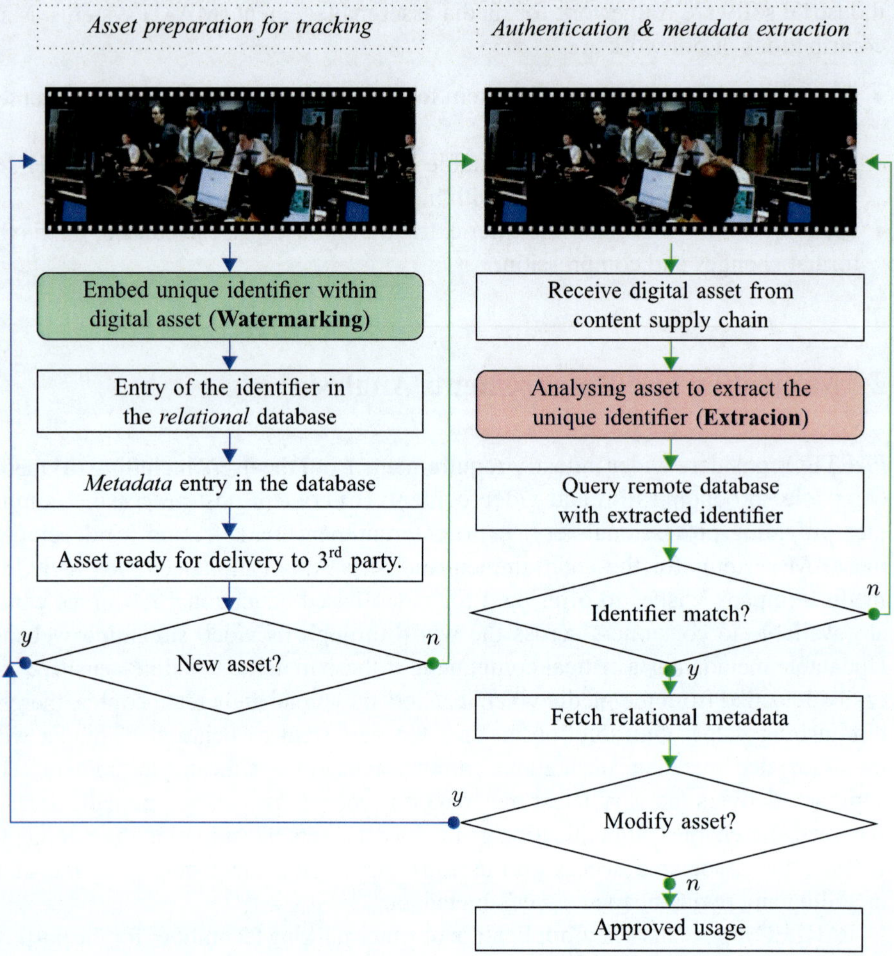

Fig. 1 PARTEC overall system diagram

ment. We also gave better emphasis on the interoperability of the proposed system with the existing framework as this is necessary for industrial adaptation. The requirements analysis is dissected in four different categories as discussed below and shown in Table 1.

Media format A solution capable of handling existing file types is important to the media industry. In PARTEC, we deal with two different categories of file types:

- Audio–video wrapper formats that multiplex video streams with optional audio streams and sub-title text and to provide one single file type (e.g., DAT, VOB, MOV, AVI, MP4, MPG, etc.).
- Video stream formats, available within the wrapped files mentioned above. These file types are encoded video bitstreams complying either proprietary (e.g., ProRes by Apple) or international standards, e.g., MPEG-2, H.264, etc.

Table 1 Types of industrial requirements

Requirements	Description
Media format	Statistical analysis of the available contents in terms of file type/wrapper format and video stream formats, i.e., compression/encoding types
Imperceptibility	Requirements for allowable quality degradation after watermark embedding
Robustness	Robustness of the algorithm that successfully preserve the watermark identifier under various compression and format changes.
Synchronisation	Ability to retrieve and identify the content which were modified such as scene editing, frame dropping, joining multiple clips, etc.

In order to achieve maximum coverage, we have analysed a statistically representative sample consisting of 168, 389 media files available from industrial partners' asset repositories, of which major wrapper types account for DAT: 43%, VOB: 21%, MOV: 15% and AVI: 13%. With respect to video stream formats, MPEG-2 occupies 47% of the total repository, followed by RAW (17%), H.264 (9%), ProRes (6%) and the remainder is made up of other formats. We have considered these statistics in designing PARTEC.

Imperceptibility Imperceptibility (visual quality) and robustness are widely considered as the two main properties vital for a good digital watermarking system. They are complimentary to each other and hence challenging to attain the right balance between them. However, in processing entertainment media, the visual quality carries significant weight to provide the highest Quality-of-Experience (QoE). Our design is heavily influenced by this requirement and ensures imperceptibility after watermark embedding.

Robustness The requirement demands two different types of robustness of the extracted watermark, (a) robustness against video format change and (b) robustness to rational compression using popular standards, i.e., MPEG-2 and H.264 here. These are needed to support a change in format or compression ratio during post-production, i.e., preparing screening quality, DVD quality or other types of content.

Synchonisation Synchronisation is one of the major issues within the video watermarking domain that is rarely addressed. In PARTEC, the synchronisation problem emerges due to media post-production including scene editing, frame dropping, combining multiple clips or inserting frames within a clip. A solution should be capable of identifying clips from multiple sources or multiple segments within a single clip. Watermark synchronisation (at least) at frame level is an essential component of our proposed solution.

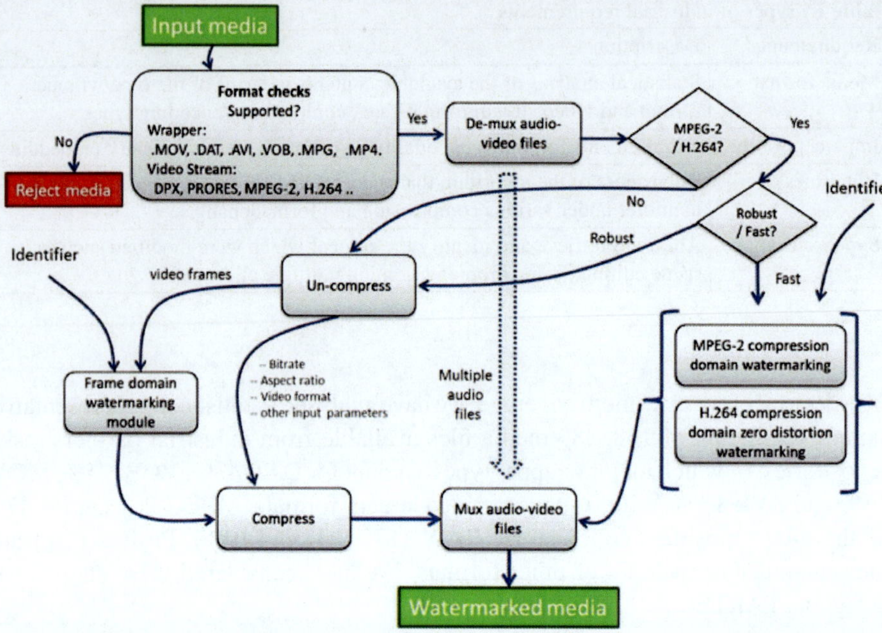

Fig. 2 Overall embedding workflow including media format handling watermark embedding

3 The System Architecture

PARTEC system architecture fulfils the industrial requirements with three major functional modules: (a) file format handling, (b) watermark embedding and extraction and (c) synchronisation. The overall flow diagram for embedding is shown in Fig. 2.

3.1 File format handling

Compatibility of the solution with existing media file types is important in the industry. As discussed earlier, two types of media files need to be handled in PARTEC: (a) audio–video (AV) wrappers and (b) video stream formats. Our strategy is to demultiplex the AV wrapper to separate the video and audio stream. While we process the video stream for watermarking, we keep the audio stream in a temporary file. After embedding, we re-multiplex the watermarked video stream with the temporarily stored audio to produce a watermarked file in the same AV format as originally received. The file format handling module also checks for supported formats, i.e., MOV, VOB, MPG, MP4, etc.

3.2 Watermarking

Based on our statistical analysis of available video formats and industrial requirements (see Sect. 2.1), evidently, we concentrate on proposing watermarking algorithms for MPEG-2 and H.264 encoded videos. However, to handle any other format, we also propose a frame domain watermarking scheme that is robust to format changes and compression. In all cases, we aim to preserve the quality of the assets closer to its original quality with minimum watermarking strengths. We also extract exact watermark sequence unlike traditional academically interesting correlation or similarity measure.

MPEG-2 compression domain watermarking We propose a new MPEG-2 compression domain watermarking module due to the widespread use of MPEG-2 video assets in the industry. Our algorithm embeds watermark on a partially decoded MPEG-2 bitstream. Firstly, the input bit stream is *entropy decoded* to produce *quantized I, P and B* frames. The motion vector (MV) data and other header information are kept separately. The partially decoded frames are then analysed and the quantised DC-DCT coefficients of the I frames are marked for watermark embedding. The least significant bits (LSB) of those DC coefficients are modified according to the incoming watermark bit. Once modified the frames are *entropy encoded* along with MV and header data to produce an MPEG-2 compatible watermarked bit stream. The overall functional block diagram of our proposed MPEG-2 compression domain watermarking module is shown in Fig. 3. Use of partial decoding within MPEG-2 flow allows us to avoid quality degradation due to re-compression. The extraction of watermark follows similar partial decoding and collects the LSB of the DC coefficient as the extracted watermarking bit.

Zero-distortion H.264 compression domain watermarking Next, we propose a zero-distortion H.264 compression domain watermarking scheme. This module embeds the watermark sequence within the Network Abstraction Layer (NAL) of the H.264-coded bitstream. The H.264 NAL standard defines 22 (/32) bits for various header information, whereas 23rd and 24th bits are *reserved* and 25th–32nd bits are unspecified. We have used 3 bits of the unspecified NAL unit bits for embedding. The bits are altered according to the incoming watermarking bit sequence: 010 and 001 for 1 and 0 watermark bits, respectively. Further, a sequence (110) is used for synchronisation after embedding of every watermark key. This allows embedding the watermark information within H.264 header without distorting the media content, and hence called *zero-distortion* watermarking. This approach provides a fast watermarking method which is robust to H.264 synchronisation attack and any H.264 file editing that preserves the header information.

Frame domain watermarking Finally, we propose a frame domain watermarking scheme that is compatible to many video formats and robust to format changes and reasonable compression. Although many joint watermarking compression domain algorithms were proposed in the literature (e.g., for MPEG-2 [5] and H.264 [9]), any compression domain watermarking scheme is vulnerable to format changes which

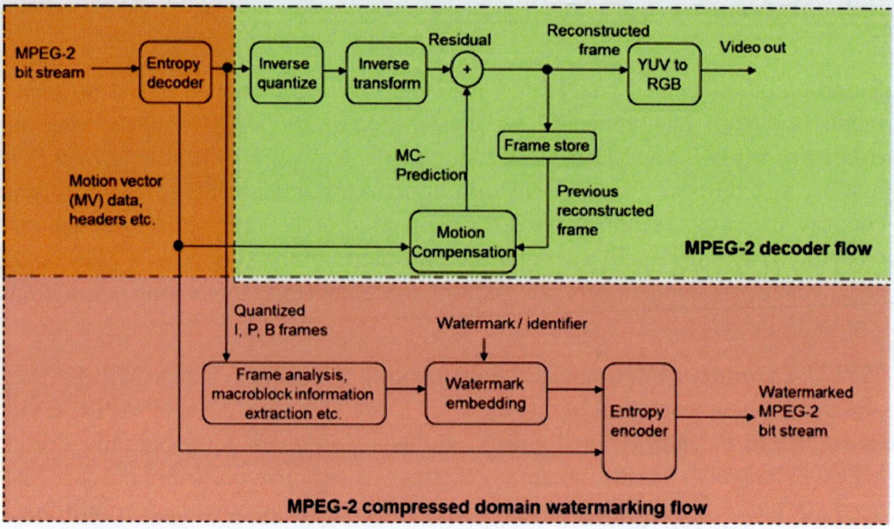

Fig. 3 Proposed MPEG-2 compression domain watermarking module

is one of the requirements in our case. Therefore, we propose a frame domain watermarking algorithm where the media bitstream is first uncompressed to individual frames, preserving all encoding-related information including information on interlacing for MPEG-2 videos. Further, we propose a new discrete wavelet transform (DWT) domain watermarking scheme as the wavelet-based algorithms demonstrated superior performances in recent literature [3].

Considering the importance of imperceptibility, we propose a novel texture based watermarking algorithm as studies [2] suggest embedding within a textured region is far less noticeable compared to embedding in homogeneous regions. Our approach also considers identifying high-frequency textured regions within the scene using DWT for watermark embedding due to DWT's dominance as a powerful tool for texture analysis [13]. DWT decomposes an image into independent frequency subbands of multiple orientations at multiple scales demonstrating details and structures.

Once wavelet decomposed the vertical and horizontal high-frequency sub-bands are divided into multiple non-overlapping blocks of size $N \times N$. The cumulative energy of the blocks for a vertical sub-band are compared with cumulative energy of the corresponding blocks in the horizontal sub-band. Depending on the incoming watermark sequence (W), the sub-band energies in pairs of blocks are modified so that the energy of one block is greater than the other and vice versa. This is achieved by modifying the coefficient values with a predefined strength parameter (α). This allows minimum distortion after embedding with reasonable robustness performance. Inverse DWT is then applied to reconstruct the watermarked frame. Finally, these frames are encoded in the same format as received using the preserved parameter information to comply with the existing infrastructure. The algorithmic description of this module is shown in Algorithm 1. A blind extractor can extract the

Input: Video stream
Format and encoding information extraction;
Decompress to frame domain;
repeat
  Wavelet transform on individual frame;
  Texture energy calculation;
  Compare vertical (V) and horizontal (H) high-frequency sub-band in blocks;
  Modify coefficients (C) iteratively: $C' = C(1 \pm \alpha.W)$ so that
  begin
    switch *watermark bit* **do**
      case *1*
        $V_{\text{Energy}} > H_{\text{Energy}}$
      endsw
      case *0*
        $V_{\text{Energy}} < H_{\text{Energy}}$
      endsw
    endsw
  end
  Inverse wavelet transform to reconstruct watermarked frames;
until *end of sequence*;
Re-compression to original video format;

Algorithm 1: Summary of frame domain watermarking module.

watermark information very quickly by comparing the block energies after wavelet decomposition of the test frame.

3.3 Synchronisation

Due to media post-processing, e.g., scene editing, video summarising for trailer preparation, often multiple clips are created from a single media source or combined to create a single clip. It is important to identify the origin of composite media files consisting of multiple clips edited together. Therefore, synchronisation at frame level or stream level for watermark identifier is necessary for this work. Our solution proposes two different synchronisation approaches to address this issue:

Frame domain unique watermarking: During the frame domain watermarking, video streams are uncompressed to frame level and unique watermark identifiers are embedded in each frame. The modularity of the unique identifier relies on the target usage. For example, frames from multiple sources can be identified by embedding one identifier in every frame for each source. Different identifiers are extracted during authentication which also indicates the temporal location of various clips. In another scenario when frame dropping or frame editing needs to be tracked, each frame from a single source requires embedding of a common identifier.

H.264 bit synchronisation: The proposed framework provides additional synchronisation for H.264 compressed domain watermarking (see Sect. 3.2). During the identifier embedding, we add synchronisation bits every time. This *self-synchronising* method allows the user to detect any cropping in the temporal domain or the presence of multiple clips (from different sources) in H.264 domain.

3.4 Overall Watermarking Work Flow

The overall flow diagram for embedding is shown in Fig. 2. During embedding, when an input media is received, it is checked for format compatibility followed by a demultiplexing of video and audio files. The audio files are kept temporarily. In the case of MPEG-2 or H.264 video stream, one can choose either a frame domain watermarking, should the requirement be for robustness. Alternatively, for fast but less robust watermarking, the MPEG-2 or H.264 compression domain watermarking scheme can be chosen. For any other formats, frame domain watermarking is recommended. As a first step of the frame domain watermarking, the video stream is first uncompressed and compression parameters, available from header information, are kept. Once the watermark embedding is done the parameters are used to re-compress the media in the similar format and quality as it was received. Finally, the re-multiplexing module combines video and audio to produce the final trackable media asset. Extraction flow is very similar to the embedding flow except, in this case, we do not need to store any temporary audio file or other parameters. Once the watermark extraction is done, an XML report is generated to collect the overall statistics of the extracted watermark/identifier. This identifier is then sent to remote database for validation and metadata extraction as shown in Fig. 1.

4 System Verification, Results and Discussions

The interoperability of our solution was tested successfully and incorporated with an industrial media asset management system. Performance of individual modules is reported here using an exemplar test sequence (Crew)[2] with a dimension of 352 × 288. Firstly, the uncompressed YUV sequence is encoded with (1) MPEG-2 compression and wrapped in a .MPG format and (2) H.264 compression and wrapped in a .MP4 format using FFmpeg to experiment with MPEG-2 compression domain and H.264 compression domain watermarking, respectively. Finally, both the formats are used to perform frame domain watermarking and tested against the requirements set out in the beginning of this paper. The proposed system is also verified with more sequences (available from the industrial partners' repository) with various media formats including, MPEG-2, H.264, ProRes, DPX and MJPEG.

In this experimental set-up for MPEG-2 watermarking, we have chosen I-frame only embedding as that is more likely robust compared to P or B frame embedding.

[2]https://media.xiph.org/video/derf/.

The H.264 watermarking does not need user-defined parameter as the module is restricted to header modification within the NAL unit to provide a zero-distortion embedding mechanism. Finally, the frame domain watermarking considers a computationally inexpensive lifting-based bi-orthogonal 5/3 wavelet kernel with one level decomposition and a block size of 4×4 (i.e., $N = 4$). The watermark consists of a sequence that represents a 64-bit binary identifier.

4.1 Results and Discussion

In verifying the compatibility with various media formats, the proposed solution is tested with various media wrappers and successfully de-multiplexed video and audio streams separately. Once the video streams are watermarked using one of the three available modules, it multiplexed the watermarked video stream with a temporarily stored audio file to output in a format same as the input one. We also compare the compatibility with video stream formats, e.g., MPEG-2 watermarking is only usable for MPEG-1 or MPEG-2 compatible streams, H.264 watermarking is usable with H.264 complied bitstream, while frame domain watermarking can be used for a wide range of video stream formats. Currently, supported video stream formats for frame domain watermarking include MPEG-2, H.264, ProRes, DPX and MJPEG. *Interlaced* videos commonly available with many existing MPEG-2 streams for broadcasting are also supported in the proposed solution. A summary of media format compatibility is shown in Table 2.

Media quality has a major influence in the design of our solution. In this work, we used two existing objective measurements: Peak Signal-to-Noise Ratio (PSNR) and Structural Similarity Index Measure (SSIM) [14], to quantify any distortion due to watermark embedding. In both cases, a higher value indicates superior performances. Generally, PSNR value more than 35dB and SSIM more than 0.9 indicate imperceptible embedding performances. The embedding distortion performances, i.e., PSNR and SSIM for MPEG-2 and frame domain watermarking are shown in Fig. 4a, b, respectively. The x-axis in each plot indicates the distortion measure and y-axis represents the frame number. In both cases, the results show PSNR values well above 35dB with an average of 46.24dB (MPEG-2) and 45.23dB (frame domain); and SSIM values above 0.9 with an average of 0.99 in both cases. The ripple effect in MPEG-2 watermarking is due to I-frame only embedding and subsequent error propagation. However, due to high PSNR/SSIM value distortions are not noticeable. It is worth noting that we do not report the embedding performance for H.264 watermarking as that does not modify the media content, hence called zero-distortion watermarking.

The robustness performances against MPEG-2 and H.264 compression are reported in Table 2. New compressed test sequences are obtained by setting the quantisation parameters in both compression standards, i.e., the quality scales (Q Scale) for MPEG-2 and the quantisation parameter (QP) for H.264. Maximum value of Q Scale = 4 and QP = 20 were set based on the current practices in DVD quality content

Table 2 Compatibility of PARTEC to industrial requirements

Compatibility ⇓	Watermarking schemes		
	MPEG-2 based	H.264 based	Frame domain (preferred solution)
Media format compatibility			
AV media wrapper (DAT, MOV, AVI, VOB, MP4)	✓	✓	✓
Video stream: MPEG-2 (including interlacing)	✓	✗	✓
Video stream: H.264	✗	✓	✓
Video stream: ProRes, DPX, M-JPEG, etc.	✗	✗	✓
Robustness against compression			
MPEG-2 compression (Q Scale = 2)	✓	✗	✓
MPEG-2 compression (Q Scale = 4)	✗	✗	✓
H.264 compression (QP = 20)	✗	✗	✓
Synchronisation			
Joining and splitting multiple clips identification	✗	✓	✓
Frame inclusion and dropping detection	✗	✗	✓

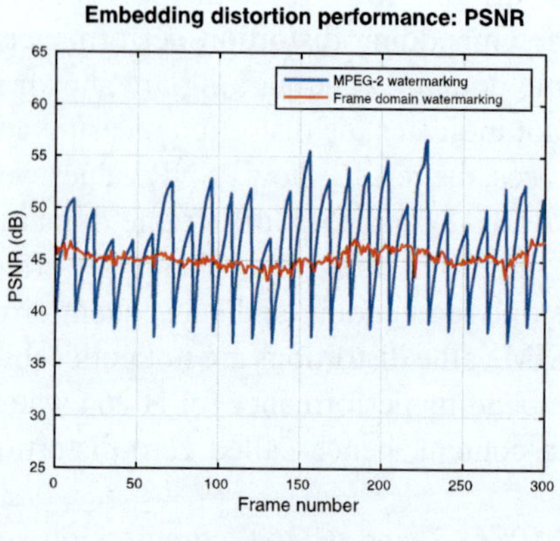

(a) PSNR plot.

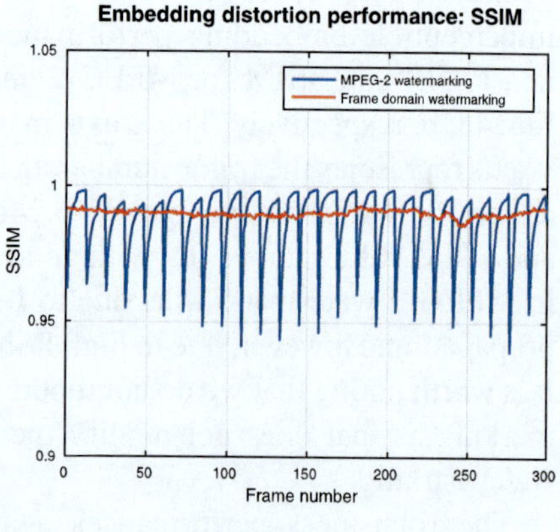

(b) SSIM plot.

Fig. 4 Embedding distortion performance of *MPEG-2 compression domain* and *frame domain* watermarking for *Crew* test sequence: **a** Average PSNR: 46.24dB (MPEG-2) and 45.23dB (frame domain) and **b** Average SSIM: 0.99 (MPEG-2) and 0.99 (frame domain) production (hence, we refrained from reporting robustness results at higher compression ratio). Extraction is applied to these compressed test sequences to retrieve the exact watermark. Our results show superior performance by the frame domain watermarking, while MPEG-2 watermarking performs within reasonable expectations. H.264 watermarking in compression tests failed in all cases as compression involves decoding and re-encoding of the media where the header information is lost. However, the latter one is the fastest to compute amongst three (frame domain being slowest) and is useful to handle synchronisation and media clip identification. We avoid reporting the robustness results for signal processing watermarking attacks, such as filtering, noise inclusion, etc., as these are not part of the requirements and hence considered outside the scope of this paper.

We report the results satisfying the requirements of synchronisation as described in Sect. 2.1. The proposed scheme can detect sources of multiple video clips (scene) joined together during scene editing or tracking multiple clips generated from a single source. The solution is also capable to identify frame dropping, frame inclusion within a sequence using frame-level watermark synchronisation (frame domain watermarking) or header-based synchronisation (H.264 watermarking). A summary of the ability to synchronise for three different watermarking modules are shown in Table 2.

Finally, we report the level of complexity for our watermarking module. H.264 only modify header information within H.264 bitstream and hence requires the least amount of computation, whereas MPEG-2 watermarking module partially decodes, watermark and re-encode the media and has reasonable complexity. The frame domain watermarking satisfies all requirements of PARTEC and robust against compression but exhibits higher computational complexity during watermark embedding. However, the extraction procedure is highly efficient providing opportunity for real-time performance.

5 Conclusions

PARTEC proposed a new application of watermarking techniques for the purpose of assets tracking and authentication. It considered wider industry requirements in both users (media or content post-processing companies) and developers (of media asset management systems) point of view. PARTEC developed the watermarking-based solution by analysing industrial requirements and produced a system that is compatible with the existing infrastructure. The requirements were dissected in four different categories, e.g., media format compatibility, imperceptibility, robustness and synchronisation. Three different watermarking algorithms were proposed in this work: (a) MPEG-2 compression domain, (b) zero-distortion H.264 compression domain and (c) wavelet-based frame domain watermarking. Experimental verification showed promising results for the target application. We conclude that our approach with the frame domain video watermarking is format agnostic and suitable to fulfil the requirements listed in this work.

Acknowledgments We acknowledge the support of Innovate UK (Project Ref: 100946).

References

1. Asikuzzaman, M., Alam, M.J., Lambert, A.J., Pickering, M.R.: Imperceptible and robust blind video watermarking using chrominance embedding: a set of approaches in the dt cwt domain. IEEE Trans. Inf. Forensics Secur. **9**(9), 1502–1517 (2014)
2. Barni, M., Bartolini, F., Piva, A.: Improved wavelet-based watermarking through pixel-wise masking. IEEE Trans. Image Process. **10**(5), 783–791 (2001). May
3. Bhowmik, D., Abhayaratne, C.: Quality scalability aware watermarking for visual content. IEEE Trans. Image Process. **25**(11), 5158–5172 (2016)
4. Bianchi, T., Piva, A.: Secure watermarking for multimedia content protection: A review of its benefits and open issues. IEEE Signal Process. Mag. **30**(2), 87–96 (2013)
5. Biswas, S., Das, S.R., Petriu, E.M.: An adaptive compressed mpeg-2 video watermarking scheme. IEEE Trans. Instrum. Meas. **54**(5), 1853–1861 (2005). Oct
6. Chan, H.T., Hwang, W.J., Cheng, C.J.: Digital hologram authentication using a hadamard-based reversible fragile watermarking algorithm. J. Display Technol. **11**(2), 193–203 (2015)
7. Cox, I.J., Miller, M.L., Bloom, J.A.: Watermarking applications and their properties. In: Proceedings. International Conference on Information Technology: Coding and Computing, pp. 6–10 (2000)
8. Fallahpour, M., Shirmohammadi, S., Semsarzadeh, M., Zhao, J.: Tampering detection in compressed digital video using watermarking. IEEE Trans. Instrum. Meas. **63**(5), 1057–1072 (2014)
9. Mansouri, A., Aznaveh, A.M., Torkamani-Azar, F., Kurugollu, F.: A low complexity video watermarking in h.264 compressed domain. IEEE Trans. Inf. Forensics Secur. **5**(4), 649–657 (2010)
10. Nezhadarya, E., Wang, Z.J., Ward, R.K.: Image quality monitoring using spread spectrum watermarking. In: Proceedings IEEE ICIP, pp. 2233–2236 (2009)
11. Stütz, T., Autrusseau, F., Uhl, A.: Non-blind structure-preserving substitution watermarking of H.264/CAVLC inter-frames. IEEE Trans. Multimed. **16**(5), 1337–1349 (2014)
12. Tsai, J.M., Chen, I.T., Huang, Y.F., Lin, C.C.: Watermarking technique for improved management of digital medical images. J. Discret. Math. Sci. Crypt. **18**(6), 785–799 (2015)
13. Ves, E., Acevedo, D., Ruedin, A., Benavent, X.: A statistical model for magnitudes and angles of wavelet frame coefficients and its application to texture retrieval. Pattern Recogn. **47**(9), 2925–2939 (2014)
14. Wang, Z., Bovik, A.C., Sheikh, H.R., Simoncelli, E.P.: Image quality assessment: from error visibility to structural similarity. IEEE Trans. Image Process. **13**, 600–612 (2004)
15. Yamada, T., Maeta, M., Mizushima, F.: Video watermark application for embedding recipient ID in real-time-encoding VoD server. J. Real-Time Image Proc. **11**(1), 211–222 (2016)
16. Zhang, H., Shu, H., Coatrieux, G., Zhu, J., Wu, Q.M.J., Zhang, Y., Zhu, H., Luo, L.: Affine legendre moment invariants for image watermarking robust to geometric distortions. IEEE Trans. Image Process. **20**(8), 2189–2199 (2011)

Author Index

J. K. Mandal et al. (eds.), *Computational Intelligence and Machine Learning*, Advances in Intelligent Systems and Computing 1276,
https://doi.org/10.1007/978-981-15-8610-1